Videocassette
Recorders
in the
Third World

COMMUNICATIONS

George Gerbner and Marsha Siefert, Editors
The Annenberg School of Communications
University of Pennsylvania, Philadelphia

Videocassette Recorders in the Third World

Douglas A. Boyd
University of Kentucky

Joseph D. Straubhaar
Michigan State University

John A. Lent
Temple University

Longman
New York & London

VIDEOCASSETTE RECORDERS IN THE THIRD WORLD

Longman Inc., 95 Church Street, White Plains, N.Y. 10601

Associated companies:
Longman Group Ltd., London
Longman Cheshire Pty., Melbourne
Longman Paul Pty., Auckland
Copp Clark Pitman, Toronto
Pitman Publishing Inc., New York

For Carole, Sandy, and Roseanne

Executive editor: Gordon T.R. Anderson
Production editor: Camilla T.K. Palmer
Text design: Steven A. Krastin
Production supervisor: Pamela Teisler

Library of Congress Cataloging-in-Publication Data
Boyd, Douglas A.
  Videocassette recorders in the Third World.
  (Longman communication books)
  Bibliography: p.
  1. Video tape recorders and recording—Developing
countries.   2. Videocassettes—Developing countries.
3. Developing countries—Popular culture.
I. Straubhaar, Joseph D.   II. Lent, John A.   III. Title.
IV. Series.   1989
PN1992.934.D44B6   1988   384.55   87–37923

ISBN 0-8013-0196-3

93 92 91 90 89 88 9 8 7 6 5 4 3 2 1

Contents

Preface

The overuse of the words explosion and revolution to describe the worldwide spread of videocassette recorders is disturbing enough to send one scurrying for refuge in Webster's. Yet, how else can the omniscient presence of the VCR be termed? Indeed, in less than a decade, the videocassette recorder and videocassette* have penetrated areas that the printing press and other information paraphernalia have not successfully reached after centuries. In some cases, governmental, religious, and other authorities, appointed or anointed to watch over sacred things such as morals, values, national interests, and power structures, are bewildered about how to control what they believe is a video menace.

Officials sometimes seem to contradict themselves when they deal with video. For example, in April 1983, a leading party newspaper in the Soviet Union sounded the alarm about watching illegal video; a year later, another major periodical did a virtual about-face, saying it was time the Soviet Union joined the Video Age. Other communist countries have been equally perplexed about the potential impact of video. Like other countries of the world, they have reacted in different ways. Strangely enough, Bulgaria was the first East European nation to set down a set of strict guidelines under the Great Anti-Video Law. Some countries have realized that if video can introduce unwanted Western messages and disseminate them, it can also transmit communist ideology and aims. The underground has learned this also, and in countries such as Poland, counter viewpoints are released in cassettes.

The same pervasiveness of video exists elsewhere. In the United States, the VCR surfaced (or was elevated by advertising) as *the* Christ-

---

* Videocassette recorder and VCR, and cassette and tape, are used interchangeably throughout this book.

mas gift of the first few years of the 1980s; in the Nordic countries, where by 1983 there were over 1 million sets, video caused what one writer termed, the "Swedish moral panic" of 1980–1984. In Africa, travelers have brought in enough sets to worry government and television authorities, one of whom described video as the "enemy in our backyards." In Fiji the government has rushed its decision to establish a national television service because of the readily available video that tourists have introduced. In 1985, 42 percent of the homes in Suva had VCRs, and 95 percent of the cassettes that were played were pirated.

As this phenomenon has developed, some institutions and scholars have taken notice and sent out teams to survey the potential and real impact of video. Much of that data is still being processed—for example, the national survey of video use in India or the assessment of the impact of video upon aboriginals in Australia's outback.

Books by Noam (1985), Gopal (1986), and the Ganleys (1986) have appeared, as have special issues or sections of periodicals such as *Intermedia, Communication Research Trends, Development Communication Report*, and *Media Development. Variety*, the United States-based entertainment weekly, recognized the phenomenon by instituting a section (and special issues) on home video; and other periodicals, such as the *Journal of Communication, Journal of Broadcasting and Electronic Media, Gazette, Index on Censorship*, and *IPI Report* have had occasional articles.

The Third World (a term used here for lack of a better one to describe most of Asia, Latin America and the Caribbean, the Arab world, and parts of Africa) has definitely figured in the video revolution. Some governments fret constantly about video, and development agencies experiment with its potential to better people's lives through more efficient information processing. Although there is literature on the various uses made of video and on individual countries' reactions to it, no compilation on the entire Third World region existed before this volume.

Videocassette Recorders in the Third World describes the video situation in the Third World, employing a geographical framework with similar thematic categories for each region: the relationship of broadcasting to video, the development and extent of video use, governance, programming, pirating, impacts, and alternative uses. In each of the three major regional treatments (Arab World, Asia, Latin America) individual countries and subregions are scrutinized; these include the Gulf states; East, Southeast, and South Asia; Venezuela, the Dominican Republic, and Nicaragua. Such a range of discussion has been provided to hedge against the tendency of books such as this one to overgeneralize data and thus to homogenize cultures, which can be unfair, if not dangerous.

The authors have conducted research specifically on video usage in these three geographical regions (and in the Caribbean) that for years have been their scholarly specialities. In fact, each author has written or edited the only book-length analysis of broadcasting in these regions. At least 60 interviews were conducted with government officials; broadcasting, film, telecommunications, and videocassette agencies and businesses; as well as with researchers studying video. In all three regions, the authors talked with personnel in video rental and sales stores and observed stocks, prices, and turnover of video hardware and software.

As with any book of this nature, there are limitations. The most serious is the omission of Africa. Although parts of North Africa are discussed in the chapter on the Arab world, the rest of the continent has not been included because we could not find an authority who has studied video usage in all of the diverse parts of Africa. The chapters, written by three individuals with different perspectives and backgrounds, may not be as consistent in style and format as some would wish, but every effort has been made to standardize them as much as possible, without losing the flavor each contributor wished to bring to his work. All three authors are responsible for writing Chapters 2 and 7; Straubhaar and Boyd wrote Chapter 1; Boyd, Chapter 3; Lent, Chapters 4 and 6, and Straubhaar, Chapter 5.

A number of institutions and people helped in gathering and processing the information for this volume. They are too numerous to list here, but they do have our sincerest gratitude. We especially thank those who helped locate material for the appendixes and those who granted permission to reprint the material. Of particular help were: Dr. Chang King-Yuh, former director of the Government Information Office of the Republic of China, and his staff; Steve Fallon; Sankaran Ramanathan; Roseanne Kueny; Luis Fernando Santoro; Regina Fesla; Henrique Geddes; Azriel Bibliowicz; Karen Ranucci; Jehan Rachty; Sami Raffoul; Yousef Shaheen; and Samir Seikaly.

Finally, the authors wish to thank Carole Boyd for the endless hours she spent on manuscript editing and indexing.

PART 1
Overviews

CHAPTER 1

The Adoption, Use and Impact of Videocassette Recorders: A Theoretical Framework

Videocassette recorders (VCRs) are undergoing an explosive growth and rapid diffusion in a number of Third World countries in Africa, Asia, the Middle East, the Caribbean, and Latin America. At least two Arab nations, Saudi Arabia and Kuwait, where VCRs are found in 80 to 90 percent of the households, far surpass even the United Kingdom and the United States. In some Third World countries, such as the Arabian Gulf states, many people are wealthy enough that the acquisition of VCRs is not surprising. But a number of less wealthy countries also have high numbers of VCRs. In many cases, the use of VCRs for entertainment, education, political expression, or commerce is having a powerful effect, both on the lives of their users and on the broadcast and other media.

This chapter will lay out a theoretical framework for analyzing the diffusion, use, and impact of VCRs in the Third World. While some examples are included here, more can be found in the regional chapters (see Part II).

Two interacting levels of causation for Third World VCR diffusion are explored. First are the systemic or contextual reasons, such as government tariff policies that make VCRs more expensive, and broadcast policies and media structures that produce certain types of media content and limit others. Second are the aggregated reasons for VCR adoption: the interests of individuals in controlling and diversifying their media consumption. These include the desires of audiences to add diversity to "television" viewing, to circumvent systemic or

3

governmental controls on content, and to gain greater personal control over informational and entertainment viewing. Such viewer needs will be contrasted with VCR costs (largely determined by government tariff policy) and individuals' income or purchasing power to consider the marginal utility of VCRs to Third World viewers with other pressing needs to be met.

VCR uses are obviously related to reasons for acquiring VCRs. However, VCR uses vary with some other factors as well, particularly cultural preferences or patterns, existing media habits, and software availability.

VCR impacts are varied, from relatively minor to heavy, depending on use and context. In some countries, VCRs have revolutionized media habits, broadcast systems, and even political systems. In others, they represent only a moderate extension of entertainment media. In yet others, VCR diffusion is so limited that no impact is discernable.

THE VCR CONTEXT: ELECTRONIC MEDIA IN THE THIRD WORLD

The diffusion of VCRs in many nations has been breathtaking in its speed. The adoption of many other inventions and innovations trails behind that of VCRs. Understanding such an enormously rapid, but very partial and idiosyncratic, diffusion requires an understanding of the prior diffusion of other electronic media. At the most obvious level, VCRs require television sets as monitors. Often, VCRs are now found in remote areas beyond the reach of broadcast television. More subtly, the use of VCRs shows a set of clear but varied relationships with broadcast television and other media: it complements them in some countries and competes with them in others.

To a great degree, the use of VCRs holds a mirror up to television, revealing a variety of faces and functions to fit what television has or has not already provided in a given place and time. Thus understanding the use of VCRs in the Third World first requires an understanding of television there, which, in turn, requires an understanding of the role of media in development.

Because of their ability to reach remote and often illiterate mass audiences, radio and television have been invested with considerable expectations and fears in most Third World nations. More than in most industrialized countries, electronic media have been seen by development theorists and planners as "magic multipliers" of educational and developmental information and by national leaders as instruments of political mobilization and control.

Mass media in the Third World have often been overemphasized in development planning and analyses of development processes. Lerner (1958) emphasized the role of media in creating empathy and awareness of the outside world and in breaking into and changing traditional ways. Rogers (1983) emphasized the ability of media to communicate innovations essential to change. Others, such as Schramm (1964), Pool (1961), and Frey (1973), also emphasized the value of media in development. All these analyses share a rather positivistic view of the media as powerful, independent, and malleable tools.

International development funding agencies, such as the United States Agency for International Development and UNESCO, as well as Third World leaders themselves, have put large amounts of funding into media development. For example, Nasser in Egypt dedicated considerable money to media development, particularly film production, television, and radio, both to promote Egyptian development and to promote his brand of Arab nationalism throughout the Middle East (Boyd, 1975).

However, beginning in the late 1960s and early 1970s, a number of researchers began to criticize the "dominant paradigm" of development thinking proposed by Lerner, Rogers, Schramm, and others. Beltran (1978), Rogers (1976, 1978), and others criticized simple receiver-oriented diffusion paradigms and highlighted the need to look at the structure, social system, and context in which those messages passed. In some contexts, development messages came to be seen as irrelevant or even counterproductive; and anticipated effects shifted "from rising expectations to rising frustrations" (Rogers, 1978). Indonesian leader Sukarno remarked in Hollywood that United States film producers were revolutionaries because they featured refrigerators in films that were shown in his country; these appliances were attractive to those in a hot climate but frustrating to the majority of Indonesians who could never afford one (McLuhan & Fiore, 1967).

This wave of criticism cast doubts on the efficacy of the media as agents of change. For instance, Whiting, Beltran, and others noted that the classic diffusion approach assumed that individuals were free to adopt new ideas and that, therefore, individuals were blamed for remaining backward, or "laggard," a term commonly used to describe those who are the last to adopt an innovation. In contrast, peasants were often described in critiques of diffusion as constrained by life circumstances, such as the lack of a surplus to use for experimental techniques, or by social structures that kept them away from supplies of credit or that kept them illiterate. In the face of such constraints, media messages for change were seen as relatively unimportant, except perhaps in creating conflict between poor peoples' aspirations and their circumstances and realities.

Also under critique was an assumption that media were, at least in some development programs, neutral or apolitical actors. In fact, critical researchers began to see media as inherently dominated by both national governments and foreign corporations. Western influence on Third World communication contexts came in for particular criticism. Pasquali (1976) and Dorfman and Mattelart (1975) looked at the role of imported media structures and media content in maintaining capitalist systems, dependency relationships, and First World ideological elements, such as consumerism, racism, and sex roles (Beltran, 1978), considered inappropriate for many Third World nations. Some in the United States, such as Schiller (1969) and Wells (1972), began directly to criticize the United States role in exporting media products, advertising, and technology to Third World nations. Varis and Nordenstreng (1973) documented the flow and criticized the role of imported media products, particularly television programs, while wire service news was examined by Boyd-Barrett (1980) and others.

Two overall paradigms for considering North-South, or First World-Third World, relations in communication emerged. First, there was the idea of cultural imperialism, which tended to make media relations a function of the capitalist world economic system. This paradigm makes mass media part of the Marxist concept of superstructure, where they reinforce and facilitate unequal and dependent economic relationships (Schiller, 1969; Dorfman & Mattelart, 1975). Another major view, media imperialism, deemphasized economic determinism and put more emphasis on unequal media and communication relationships per se (Boyd-Barrett, 1980; Lee, 1980). Both viewpoints considered VCRs to be another means of widening the unequal and unbalanced flow of film and television from the First World to the Third World.

Another wave of research looked at the influence on the Third World of imported broadcasting models from the industrialized nations. It focused particularly on the British model (BBC public service, license fee-based systems), the American (tending to commercial financing and a mix of libertarian and social responsibility ethics), and the French (more state control and mixed government-commercial financing). Such studies were done by Tunstall (1977), Read (1976), and Katz and Wedell (1977). Later work by Lee (1980) and others began also to look at the USSR (government-controlled media as an instrument of development) as a signicant model for a number of Third World countries.

In many cases, models from abroad were modified substantially, as was the American model in Brazil, for example (Mattos, 1984; Straubhaar, 1984). Other models were copied less extensively, such as Nasserite "mobilization" systems in the Arab world. In all, a number of diverging systems appeared. Third World systems can now be found to fit all 'four theories of the press' (Siebert, Peterson & Schramm, 1956), as well as the more recent "development journalism" model (McPhail, 1981; Rosenblum, 1979). The broadcast model chosen in a given country is quite important for predicting VCR diffusion and VCR impact. Several Third World broadcasting tendencies are particularly significant for VCR diffusion. Those include the issues of centralized versus individual control over content, education versus entertainment in content, and national content promotion versus desires for free flow of content across borders. Each of these issues has systemic, structural, policy, and audience-demand aspects, which will be related to VCR diffusion here. They also set up a base for later examining the impacts of VCRs on broadcasting systems and society, which will be taken up at the end of this chapter.

THE RAPID DIFFUSION OF VCRS

The spread of VCRs has been recent, explosive, and rapid. Preliminary empirical research (both market research and historical and structural analysis) has identified several factors involved in VCR diffusion: price, variations in price because of government restrictions, income and income distribution, the content of broadcast television, the diversity of entertainment media available, and, for minority cultures, the degree to which the languages and cultures of audiences are represented in the boradcast system.

The first three factors are linked. If prices are low, a much broader audience, including lower income groups, can have access to VCRs. This is increasingly the case as VCR prices have continued to decline from a world average of U.S.$400–$500 in the early 1980s to U.S.$200–$300 in 1986. Where prices remain high, it is usually because of attempts by governments to discourage VCR acquisition by imposing tariffs or import barriers. If local manufacture is required, as in Brazil, prices increase as much as 200 or 300 percent. Such barriers are widely violated by smuggling, but barriers increase the price and restrict access to VCRs (Ganley & Ganley, 1986; Heeter,

1984). In several countries, a significant proportion of VCRs are smuggled in (Ganley & Ganley, 1986). Even where VCRs are manufactured, as in Brazil, local supplies may not meet demand or may be more expensive. In Brazil, roughly two-thirds of existing VCRs are estimated to have been brought in from outside, primarily by smugglers ("Videocassette no Brasil," 1986).

Even when prices have not been artificially increased by tariffs or barriers, the marginal cost of a VCR—i.e., the cost relative to buying power and other pressing needs—is still very high for an individual, family, or even a village in most Third World countries. That is, a much larger proportion of income will be required to buy a VCR than in richer nations. Therefore, the marginal utility of the VCR must also be correspondingly higher. The prospective VCR buyer in most Third World areas does not make a casual purchase; the machine must offer something sufficient to make it prized above competing goods, such as washing machines, refrigerators, or vehicles. One of the most impressive aspects of the VCR boom is that the video machines do surpass these alternatives for many. For that to happen, the VCR must add greatly to the gratification, status, or quality of life of the viewer. When the VCR acquires such value and importance, the impact of the machine on media habits, social life, and society is correspondingly greater.

According to correlational studies carried out by Heeter (1984), Lin (1985), and Straubhaar and Lin (in press), one of the variables most significantly associated with VCR penetration is the diversity of what is available on broadcast and cable television. Both the number of broadast or cable channels, per se, and more content-oriented measures of diversity are correlated with VCR penetration, positively in some circumstances and negatively in others.

In the Third World, as well as in industrialized nations, VCR penetration must be put in the larger perspective of video media competition (Noam, 1985; Lhoest, 1985). In some wealthier countries, such as the United States or Britain, or in a few of the more media-rich developing countries, VCRs seem to be used primarily for time shifting (recording programs off-air for later playback). In wealthier countries, VCR penetration was positively correlated with measures of broadcast diversity, number of TV channels, and cable-TV availability (Straubhaar & Lin, in press). These two relationships make intuitive sense. Where a relatively rich video environment exists, and where individuals have the buying power to make a relatively casual purchase of a VCR, then VCRs can and will be acquired for convenience, building up off-air libraries, and effectively enabling viewers to watch two programs broadcast at the same time. As noted later, this presages an impact on broadcasting, although a relatively minor one. Advertisers, however, may suffer because of "zipping" (fast-forwarding through commercials), "zapping" (the use of "pause" or "stop" controls to eliminate commercials from recordings), or knowing neither who is watching advertisements nor when with simple time shifting.

However, in most Third World countries, VCRs spread fastest where television viewers feel they are not well served by what is broadcast—in short, where the marginal utility of VCRs for adding diversity to viewers' video diet is very high. In poorer countries, VCR use is negatively correlated with measures of TV diversity (Straubhaar & Lin, in press). The explanation seems to be that in those Third World countries with relatively ample or diverse television broadcasts, but where individual or group resources must be more carefully guarded because of lower incomes, the marginal utility of VCRs is lower. Similarly, VCR penetration is higher in Third World countries that have only one broadcast channel, that have few entertainment channels, or that strictly control entertainment and information. In these countries, the audience need for entertainment or diversity makes the acquisition of a VCR a necessity. It is also clear that the use of VCRs in such circumstances, where audiences are extensively avoiding broadcast television content, creates a much higher impact than in countries where VCRs are primarily used for time shifting.

In either case, VCRs offer greater personal control over what is watched. VCRs enable people to personalize their video viewing and better meet their own perceived needs or interests. To some degree, this is simply a Third World manifestation of what has long been predicted in communication research; a more active audience that seeks to gratify its own needs with selective media exposure (Rosengren, Wenner, & Palmgreen, 1985).

The personalization of video consumption by means of VCRs enables wider variety among the programs that an individual, family, or larger group may choose to watch. Still, some patterns do emerge. Above all, VCRs seem to be predominantly used to add entertainment material where relatively little entertainment is available on broadcast television or where certain kinds of entertainment are not broadcast. In analyzing why audiences seek entertainment, i.e., the utility it has for them, Atkin (1985, p. 64) noted that there are both positive drives for content that is enjoyable, and negative selection to avoid tension, boredom, and loneliness, and to escape from problems.

This profile of VCR diffusion and use matches up relatively well with the patterns and context of Third World media. Several major relationships between the diffusion, use, and impact of VCRs and Third World media patterns deserve further discussion.

CENTRALIZATION, BROADCASTING, AND VCRS

At least until recently, in most approaches, the use of electronic media for development was seen as properly, or perhaps inevitably, centralized. Ambitious experiments, such as India's Satellite Instructional Television Experiment (S.I.T.E.), relied on centralized production and distribution of messages (Mody, 1979). Although evaluations showed that one of the more successful parts of the overall S.I.T.E. project was a regional production and broadcast operation in Kheda, that part was also the most politically and bureaucratically problematic, and has, in fact, been shut down (Chiruvolu, 1986). The great political advantage of S.I.T.E. was, it seems, the opportunity offered to Indian national leaders to communicate more directly with remote parts of the country not previously reached by mass communication.

As with the Kheda project, newer theories and fieldwork in development communication show that decentralized media are more effective in reaching and holding audiences (Berrigan, 1979; Chiruvolu, 1986). Other work with community radio stations in the Caribbean, Latin America, and Africa indicates that Third World audiences react very favorably to media that provide greater local community feedback, or interaction, or formative input (Jamison & McAnany, 1976; White, 1976; Mody, 1986).

Nevertheless, the temptation for Third World leaders to centralize and control broadcasting is very strong. An overview by Head (1985) placed most Third World systems in the general category of "authoritarian," or government-dominated. While seeking the totalitarian control exercised by the USSR, relatively few are willing to entrust the media themselves with final responsibility for content, as the libertarian or social responsibility models would prescribe.

The detailed study of Third World broadcasting conducted by Katz and Wedell (1977) indicated that despite an initial tendency to copy United States, British, or French broadcast models, modifications, particularly in the area of political control, had been widespread. In particular, they noted that the British ideal of BBC-style independent corporations had not fared well in the often unstable politics of many former British colonies. Even such relatively democratic states as Nigeria and India had put broadcasting either into government ministries or under much closer federal control. Similarly, Latin American countries that had used the United States model of private commercial systems often showed far more assertion of government control over content, both reactively in terms of censorship and proactively in terms of guiding content (Alisky, 1983).

One of the effects VCRs may have is to challenge this prevailing system of government controls on television. (Audiocassette recorders are doing much the same thing in radio broadcasting.) VCRs circumvent the use of television for both development and political control by bringing de facto media decentralization. They put the control of programming in the hands of the audience. What VCR owners choose to watch (and what is most available to them) seems likely to circumvent and frustrate both development programs' intentions to educate them and leaders' desires to control the information content available to them. This extends to family control as well; a survey of students in Saudi Arabia showed that many of them used VCRs to watch programs they felt their parents would disapprove of (al-Oofy, 1986).

VCRS AND BROADCAST TV DIVERSITY

Centralized broadcasting systems create structural conditions that lead to a desire for greater individual control via VCRs or other means. A major reason is the control of content inherent in mass media, which is more explicit and clear in many Third World countries, particularly in broadcasting. In most Third World countries outside Latin America, most broadcast media systems are operated or controlled by governments. These systems were created with goals consistent with the British public service concept, the French notion of cultural nationalism and political control, the development journalism concept of cooperation with development programs, or Soviet- or Chinese-style total media involvement with government programs. All of these goals exclude certain kinds of television content. In particular, few of these systems and patterns emphasize television entertainment.

Individual or small-group control of video consumption via VCRs enables a wider variety of programs that an individual, family, or larger group may choose to watch. In Indonesia and Malaysia, for example, "the public seeks video because of the limited fare of the state television service" (Lent, 1985). Overall, quantitative studies of national patterns of VCR viewing suggest that VCRs are predominantly used to add entertainment material in areas where relatively little entertainment is available on broadcast television or where certain kinds of entertainment are not broadcast (Straubhaar & Lin, in press).

In places such as Saudi Arabia, where television has been highly didactic, stressing religion, education, cultural improvement, or political persuasion, almost any entertainment seems to be snapped up for VCR consumption. What tends to be most readily available are feature films, followed by television series taped off-air (Boyd, 1985; Lent, 1985; Ogan, 1985; Straubhaar, 1986). Most often these are from the United States, followed by European, or incresingly, local or regional productions (Varis, 1984). In Saudi Arabia, series and films imported from both the United States and Egypt, the primary regional producer in Arabic, are popular.

Even in more entertainment-oriented television broadcast systems, specific kinds of entertainment are frequently heavily edited, downplayed, or prohibited. Perhaps the most visible and controversial example is pornography or, more generally, sexual content in both television series and films. Pornography and even much less explicit sexual content is particularly worrisome to authorities in Islamic countries (Boyd, 1985; Lent, 1985). Various case studies show that pornography frequently dominates VCR viewing, particularly in areas where the VCR is a new phenomenon. Even so, the novelty effect wears off and the frequency of pornography viewing often declines ("Video cassette recorders," 1983), particularly if children are watching when parents are not at home. In Latin America, for instance, estimates of the proportion of VCR use devoted to pornography vary from over 50 percent (O'Sullivan, 1985) to under 10 percent (interviews, Caracas, 1985, and Brazil, 1986).

Also controversial are news and political information on videotape. A study by Ganley and Ganley (1986) focuses on the potential of VCRs for allowing politically challenging or subversive material to bypass national censorship and counter national points of view or propaganda. They find the political potential of VCRs to be considerable. The main case to date is the Ayatollah Khomeini's use of audiocassettes, photocopying, and VCRs to spread opposition to the Shah of Iran and create support for his Islamic revolution. The Philippine opposition movement, led by Corazon Aquino, appears to have used videotapes of foreign coverage of her husband's assassination to accentuate indignation about the slaying and opposition to Ferdinand Marcos (Lent, 1985).

Other less controversial genres of material are also sometimes limited by broadcast policy. A number of countries limit showings of some *Rambo*-type American adventure films and series that are considered too violent but remain popular with sizeable parts of the audience. For example, in Brazil, which had fairly minimal controls over imported programs, the Minister of Communication, Quandt de Oliveira, under the most conservative of the Brazilian military regimes, pushed hard in 1972–1973 for a reduction in imported United States programs because of stated concerns over the effect of violent programs on Brazilian children and Brazilian culture (Straubhaar, 1984).

Indian movies are similarly popular but proscribed in Pakistan because of conflict between the two nations and, in some cases, because of sexual content (Lent, 1985).

VCRs also play a visible role in supplying minority audiences with programs that are not provided by television broadcasts. In many nations, only one or two channels exist, with fairly narrow goals. In contrast, many nations have very diverse ethnic and linguistic minorities who demand television programs targeted to their needs. Furthermore, there are interest groups, ranging from political opposition groups to martial artists and bird-watchers, who want specialized material.

Perhaps most significantly in the Third World, VCRs can help ethnic or linguistic minorities find material that uses their languages and that caters to their cultures. For example, Chinese communities in several Southeast Asian nations are not directly served by broadcast television, which as a matter of national policy emphasizes the dominant languages of the nations. In Malaysia, for instance, broadcast authorities acknowledge that they are losing the Chinese audience but are reluctant to modify their language policy and goals (Lent, 1985; Tan, 1987). English-speaking travelers are often catered to in hotels by in-house cable systems fed by VCRs with videotapes flown in from the United States or United Kingdom, while a number of ethnic groups in the United States and United Kingdom use VCRs to watch television and films from home (DoBrow, 1986).

PROGRAM FLOWS AND VCRS

A great deal of the New World Information Order debate has centered on the flow of films, recorded music, news agency reports, television programs, and other cultural products from country to country. In many cases, the impact of imported media on national, regional or local culture has been shown to be at least somewhat worrisome to leaders (Beltran, 1978; Boyd, 1985; Straubhaar, 1984). In several countries where VCRs are widespread, leaders are now concerned that VCRs will provide another channel for program imports. One of the goals of broadcast policy in a number of countries has been to limit the inflow of foreign culture on film, music, and video, and to promote national production (McBride commission, 1980). Even in Great Britain and Canada, imported material is limited to a certain proportion of broadcast time. However, as with the examples above, this may pit government policy against the interests of individual viewers, who might prefer to watch imported material.

This conflict is accentuated as VCRs give individuals or ethnic minorities the ability selectively to import television programs, music videos, or films from other countries. As noted above, the importation of Chinese films into Malaysia and Indian films into Pakistan has already reached sufficiently significant levels to concern authorities there (Lent, 1985). Other such specific regional video flows are likely to increase—for instance, from Egypt into more conservative Arabic-speaking nations (Boyd, 1985).

The main TV flow issue is the possibility that VCRs will add to the already pronounced flow of Anglo-American film, television, and music video material in English to Third World countries. Surveys of video rental shops by the authors in several countries in Latin America, the Middle East, and Asia show that American movies (and television shows in the Middle East and Asia) dominate much of what is available for rent or exchange in shops or clubs. Much of the material is dubbed, taken from off-air broadcasts, but much is also either subtitled or available in English only, particularly very recent films.

There seem to be several reasons for this use of VCRs for the continued outflow of American programming. First, from the early days of cinema American movies have set a pattern for what is considered entertaining in many countries (Guback, Varis, et al., 1984; Read, 1976; Tunstall, 1977). Second, the flow of films, video, and television programs between countries has been dominated by Anglo-American products. Therefore, after years of avid exporting by the United States, the stock of available films for either legal or pirated distribution via VCR is highly dominated by American products. So when individuals look for entertainment to supplement viewing, the main options they may find, because of existing film and television distribution patterns, are American films pirated from theatrical distribution or series recorded off-air. Third, the fact that many Third World countries are trying actively to reduce American and other media imports may repress an existing demand for imported entertainment television and film that has built up over the years. If an audience has developed a taste for American films and television, VCRs may give them a vehicle for ignoring official restraints.

VCRS AND THIRD WORLD GOVERNMENT MEASURES

A last major area of Third World development policies related to VCR diffusion is government policy that affects prices and the availability of imported VCRs. Except in South Korea, and, more recently, Brazil and

India, VCRs are imported goods in the Third World. As Table 1.1 shows, as of 1984 VCR prices in a number of countries averaged about U.S.$400. That seems to have been the natural market price, except where extremely difficult trade conditions or government import policies raised prices.

Some governments deliberately set high taxes or tariffs on VCRs to restrict their importation. Other governments simply prohibit the importation of VCRs through legal channels for a variety of political and economic motives.

A number of governments, including First World nations such as the USSR and France, have tried to restrict VCR imports out of fear of the kinds of cultural and political effects mentioned above. These governments try either to eliminate or radically restrict the importation of VCRs. However, flat prohibitions or restricted quotas have not worked very well. In many countries, VCRs are smuggled in by the thousands (Ganley & Ganley, 1986). Such measures do raise prices, though, since the smuggled goods are usually more expensive. In some cases, such as India, governments have eventually stopped trying to keep VCRs out for political or cultural reasons and have moved to more economic concerns about reducing imports.

A number of countries have concentrated more on VCRs as an economic issue. Some, such as Brazil or India, want to create domestic VCR-manufacturing industries in order to reduce or provide substitutes for imported machines. This follows from a general tendency, particularly among the larger Third World nations, to follow a policy of import-substitution industrialization. Governments that follow this economics-based policy do not object to the consumption of goods per se—even luxury goods—but want to avoid foreign exchange problems and to create employment and industrial and technological bases in their own countries by substituting nationally manufactured goods for imports. In any case, such a policy tends to raise prices. Cheaper imports are either blocked or heavily taxed, while locally manufactured goods, portected by such policies, are more expensive.

For many smaller countries, the idea of producing or even assembling VCRs from kits of imported parts may not be feasible. Still, many of these countries see VCRs as a luxury consumer item and tax them accordingly. This policy also raises prices.

Running counter to either import-substitution or tariff policies on VCRs is the overwhelming demand for them in a number of countries. Still, as we have seen, the demand for VCRs is not inflexible and depends largely on how satisfied audiences are with other media. If demand is not, in fact, extremely strong, then restrictions that raise prices can reduce VCR diffusion considerably. In Brazil, for example,

TABLE 1.1 VCR PENETRATION IN 1984

Country	% of TV Homes with VCRs[a]	Total[b]	No. TV Channels	Average VCR Price
Algeria	*	*	1	—
Argentina	5##	1##	4	850
Bahrain	79##	—	2	400
Bangladesh	1	*	1	1,000
Barbados	2	*	1	—
Brazil	11##	2##	5	1,000
Cameroon	—	*	0	2,300
Chile	4	1	5	800
China (PRC)	*	*	1	1,900
Colombia	25	2	2	1,950
Dominican Rep.	—	*	5	800
El Salvador	2	*	5	500
Egypt	4##	*	2	2,500
Ethiopia	8	*	1	1,100
Fiji	—	6	0	450
Ghana	8	*	1	700
Honduras	3	*	2	500
Hong Kong	27	8	4	400
India	29	*	1	800
Indonesia	6	*	1	—
Iran	13	1	1	—
Iraq	24	1	1	—
Israel	56	9	2	—
Jamaica	8	1	1	2,450
Jordan	30##	3##	2	2,100
Kenya	10	2	1	1,500
Kuwait	88##	55##	2	350
Lebanon	14	1	2	450
Liberia	5	*	1	1,000
Libya	84#	6#	2	—
Malaysia	30	3	3	400
Mali	*	*	0	—
Mexico	5	1	4	550
Nigeria	2	*	0.5	1,000
Oman	75	55	1	400
Pakistan	15	*	1	—
Panama	10	1	5	400
Peru	11	1	5	—
Philippines	14	1	3	400

(TABLE 1.1 Continued)

Qatar	77##	25##	1	400
Saudi Arabia	75	—	2	400
Singapore	25	14	1	400
South Africa	20	2	1	—
South Korea	7#	1#	4	400
Sri Lanka	5	*	2	
Sudan	*	*	2	2,000
Syria	10##	1	1	—
Tanzania	*	*	0	2,000
Taiwan	19	8	3	700
Thailand	9	*	4	—
Trinidad	47	9	1	500
United Arab E.	80	7	2	400
Uruguay	—	—	4	1,000
Venezuela	14	2	3	500
Yemen (North)	25##	3	1	650
Zaire	50+	1	2	1,000
Zambia	*	*	1	—

[a] The proportion of households with TV and VCRs.
[b] The proportion of VCRs to the total population of the country.
* Indicates less than 0.5 of 1 percent.
# These data are from 1983.
These data are from 1986; all others are from 1984.

despite a higher GNP per capita than Colombia, VCR penetration is much lower (Table 1.1). This discrepancy seems to result from the greater diversity and higher quality of programs on broadcast television, according to interviews by one of the authors (Straubhaar in Brazil, 1986). However, to give economic factors their due, a revived economy in Brazil may have led to greatly increased purchase of VCRs in 1986, not so much to substitute for broadcasting as to let owners create new content with cameras or meet special interests in sports, health, or art films that are not released in the country ("Videocassete no Brasil," 1986).

VCR OWNERSHIP PATTERNS

Information about ownership of, or access to, VCRs is still incomplete. Governments are inefficient in collecting import and sales data. Even manufacturers' sales data are often misleading. Part of the problem is the widespread purchase of smuggled or black market VCRs. Ganley and Ganley (1986) reported both the importance of the black market to many economies and the particularly extensive sales of VCRs on the black market in countries as diverse as Burma, the People's Republic of China, Mexico, Nigeria, and the USSR. If, for example, manufacturers' sales figures were accurate for Panama, a major smuggling point, almost all residents there would have a VCR (Ganley & Ganley, 1986). In the Arabian Gulf, as much as 50 percent of imported machines are taken home by expatriate workers or smuggled by sea or land to other countries.

In some countries, market surveys have estimated VCR ownership on the basis of samples. In most other Third World countries, expert estimates are probably more accurate than official government or sales data.

Table 1.1 gives VCR penetration levels as of 1984, the last year for which broadly comparable data were available. Newer data were used when available.

VCR SOFTWARE SALES AND PIRACY

Given the demand, a number of new businesses have started to supply video tapes to very diverse audiences. Most tapes—75 to 80 percent in most countries—are *pirated* (i.e., copied illegally without compensation to copyright holders). Therefore, the copying and sales of these tapes tend to operate as cottage industries in the Third World economic underground. Even countries such as the USSR and France have been unable to stop the illegal importation of VCRs, much less tapes of undesirable or pirated material, which can be easily carried in a pocket (Lhoest, 1983; Smale, 1983). Even the most authoritarian governments in developing countries have no greater control over programs deemed contrary to national interests.

The availability of forbidden content is one motive for piracy on video. Another is purely economic: pirated tapes are cheaper than legally imported tapes in all of the rental shops that the authors visited on several continents. Depending on price, availability, and controls over what films are released in the theaters, the option to see pirated films seems to be a significant incentive for buying a VCR and has been one of its principle economic effects. Both national and international film industries have been affected by piracy, since the viewing of pirated tapes competes directly with theaters. Furthermore, if a substantial percentage of the audience has already seen a film on VCR, even the television market for that film may be lost. For example,

Jordanian Television (JTV) approached the American Broadcasting Company (ABC) for permission to show the 1983 television movie *The Day After*. After returning to Jordan, the JTV program director discovered that a substantial portion of the population had already seen pirated versions of the film in Amman, so he notified ABC that JTV was no longer interested in the film (M. Kheir, personal correspondence, Amman, Jordan, January 14, 1984).

Most VCR software piracy involves the importation of entertainment programming from other countries ("Video cassette recorders," 1983; Ganley & Ganley, 1986). The Motion Picture Export Association of America (MPEAA) estimated in 1985 that its members had lost U.S.$700,000,000 in potential theater ticket receipts because viewers in a number of countries had watched United States movies on videotape instead of in movie houses. In just one country, Venezuela, where VCR penetration of households is 18 to 20 percent, losses by MPEAA distributors were estimated at U.S.$20,000,000 in 1981 alone ("'Legal' homevid," 1985). The United States companies belonging to the MPEAA are not the only ones in the international film industry to have been hurt. As early as 1983, it was estimated that video piracy had cut as much as 80 percent of the earnings of the Indian film industry from distribution in Southeast Asia, the Middle East, and Africa ("India's dream merchants," 1983).

In some countries, piracy is becoming a domestic issue as well. Several Third World countries have major film or television industries that produce entertainment. In those countries, piracy via VCRs threatens these national film industries. In India and Nigeria, for example, VCRs have posed a serious challenge to national film distribution. "Video parlors" show pirated films at cheaper prices than movie houses. The Indian parlors frequently show films before they are released in rural or suburban areas (Ninan & Singh, 1983). For example, the Minerva Theater in Jodhpur was scheduled for reopening with the film *Jani Dost*. The manager said, however, "after 800 to 1,000 people have been seeing the film daily at these video-cafés (eight near the theater), we really don't know what sort of market will be left" ("Video boom," 1983). The film industry in Bombay and other cities has felt quite threatened and has appealed to both the legislature and the courts for protection from VCR piracy.

Governments also experience the effects of VCR piracy. Several Third World countries finance their film industries by taxes on theater admissions. Even by 1981, the Philippine government had estimated a loss of U.S.$2.1 million because of a 30 percent drop in cinema admissions ("Homevid is blamed," 1981).

VCRS AND ALTERNATIVE PRODUCTION

In some countries, significant use of VCRs is being made to create and view *alternative material*—i.e., material not previously shown on film or broadcast publicly. In some cases, as in Brazil, this takes the form of sports programs, how-to information, and custom-made programs about families, childbirth, and weddings ("Videocassete no Brasil," 1986). In India, similarly, the videotaping of family events (especially weddings), business meetings, and community affairs is becoming both a hobby for some and a business for others (*India Today*, June 30, 1983). This practice is spawning a second wave of VCR adoption. For example, in Brazil VCR adoption remained limited when the primary use was only for rental of pirated films, but it has grown enormously in the last year as other uses have been publicized ("Videocassete no Brasil," 1986).

In Brazil and elsewhere in Latin America, where most broadcasting is already entertainment oriented, VCRs are also being used by community groups, unions, political parties, and even guerrilla groups, to spread alternative political and social messages. The extent of such use is difficult to gauge; hence its significance in VCR adoption and impact is somewhat unclear, except that it creates a completely new set of uses for the equipment, with a substantially new set of users (interview, Karen Ranucci, Philadelphia, October, 1986; "Democracy in Communication," 1986).

However, the potential impact of such alternative political communication via VCRs is clear. Groups and individuals may spread completely independent, even subversive, messages that would be censored by broadcasters or government officials from regular television. Following are two examples from a sample tape assembled by Democracy in Communication. (This descriptive text and other examples are contained in the brochure "Democracy in Communication," reprinted in Appendix 5.A.)

"Amas de Casa" ("Housewives") (videotape, 5 minutes, 1984) was produced by the Colectivo Cine Mujer (Women's Film Collective), a group of women filmmakers who joined together in 1978 to produce programs which would be relevant to their interests as Mexican women. In this video tape project, women from one of the city's many cardboard villages act out a familiar situation. A neighbor, late with her rent payments, is being evicted. The entire neighborhood bands together to stand behind her in defiance of the court's eviction order. This tape has been used as an organizing tool by the housewives' union in this neighborhood. The process of making the tape helps the women gain the confidence they need to face this situation again, in real life. Today the Colectivo Cine Mujer no longer exists. Rather, each of the women [is] working separately in some aspect of cinema production.

"Algo de Ti" ("Something of You") (music video, 5 minutes, 1985) won second prize in the 1984 Panamanian Music Video Festival. It represents the integration of music with social statement. The song is a love song, but the video speaks of the horrors of living under a military government. A man is arrested with no explanation given. Surrealistic images imply that society lends its complicity through its silence about such acts of violence. It was made by Luis Franco and Sergio Cambefort of Boa Productions. They work in commercial video production and do their own work on the side.

One problem for producers of such material is how to gain an audience. Some groups produce videotapes primarily for their own use as training tools, or simply for the empowerment and acting out of conflicts that come from making such tapes, as the first example above shows. Others, however, clearly want to reach a wider audience. In most cases, that audience is local, regional, or at least national; but in some cases, groups are now aiming for international distribution and effects, as the following Democracy in Communication description of video work by the El Salvadoran guerrillas, the Farabundo Martí Liberation Front, indicates.

> The guerillas feel they must communicate directly with the people of El Salvador to combat the government's control of the media and relate their side of the conflict to the populace. Their primary activity is in the form of daily radio broadcasts. In order to operate unde-tected by the military, their transponder is portable and constantly moving.
> Their video work serves audiences both inside and outside the country. Their tapes are taken to mountain villages and shown to peasants as a way of keeping them in touch with the war that is being fought in other parts of the country. Their tapes are also shown in foreign countries as a means of gaining international support for their cause.

In some countries, the Roman Catholic Church has begun to serve as a focal point for producing alternative videos, primarily at the level of the parish or "base community." The church structure, particularly its film and cultural organizations, then also serves as a channel for exchanging videos. Other organizations are also taking up that role. In Brazil, for example, there is a national association of independent community video producers supported, at least from 1984 to 1986, by the liberal government of the state of São Paulo, which furnished office space, salaries, and organizational help to the independent association, and created its own plan to circulate videos to regional libraries (Interviews, Santoro and Milanesi, 1986) (See Appendix 1.A, for an example of the use of video in African villages).

THE INSTITUTIONAL AND SOCIETAL IMPACT OF VCRS

In general, the major impact of VCRs to date has been on Third World broadcasters. In several countries, the audience for broadcast television has been decimated. In countries with extremely high VCR penetration, such as Saudi Arabia or Kuwait, broadcasters really do not know who is watching what and when. Even a number of countries with much lesser VCR penetration are changing the broadcast system to react to or anticipate VCR effects. For example, Malaysia has now created a second channel with more entertainment and more minority language programming to try to reclaim minority audiences who are using VCRs as their "television" channel of choice (Tan, 1986).

Third World film producers have also been powerfully affected, as the experience of India, noted above, shows. Even when audiences are largely limited to elites and the middle class, as in India, the effect on theater gate receipts for film distributors can be very considerable.

International film distributors have also been hit rather hard by the impact of VCRs. If an increasingly major proportion of film viewing is being done on VCRs (25 to 30 percent in several countries even in 1983–1984, then a great deal of revenue is lost to both national and international producers and distributors. An MPEAA attorney estimates their revenue lost to VCR piracy at close to U.S.$1 billion (Gandelman, 1985).

Impacts on society are hard to assess yet, but potentially broad. Children increasingly have access to adult programming. Evidence from countries as diverse as Saudi Arabia (al-Oofy, 1986) and Venezuela (interviews, 1985) indicates that adolescents are active and independent in using VCRs to acquire and see material that their parents would probably not approve of, following in the patterns already shown in the United States (Greenberg & Heeter, 1987). And VCRs allow viewing of socially taboo material, such as pornography or violence, and facilitate bringing in programs from other countries, even when governments would like to restrict or censor them.

In summary, there are a series of interacting systemic reasons why individuals in many Third World countries might supplement existing broadcast or film fare. Centralized broadcast structures work against individual, ethnic, and regional variety in audience interests. Didactic, propagandistic, or educational system goals frequently limit access to content, particularly to entertainment, that audiences may desire. The domination of broadcast news and commentary by a few political points of view may lead those with dissenting or alternative views to seek out other information or commentaries. In all of these cases, by using VCRs to obtain the "television" programming that they want, viewers subvert or circumvent government plans to limit or control television content to achieve specific goals. In this sense, VCRs may indeed be a revolutionary medium in many states and societies.

APPENDIX

1.A: The Use of Video in the Village
by Dembele Sata Djire

Much of the discussion in this book examines the home video tape recorder in the context of entertainment. There are many reasons for acquiring a VCR, but a great advantage of this home appliance is its potential to liberate the owner from what is often limited programming on television. The history of media to promote national development is well documented, but efforts to use video tape recorders for this purpose are still new. One such effort is described by Dembele Sata Djire, Chief, Division for the Promotion of Women, National Department for Functional Literacy and Applied Linguistics, Republic of Mali. The paper was presented in February 1983 at the Tenth Anniversary Colloquium of the United Nations Development Forum.

Technology cannot do anything on its own. It has to be selected to address identified problems; and then it has to be applied by people to solve them. In Mali, we have chosen video as a tool, an instrument for promoting rural development, particularly to improve the conditions of women.

The Republic of Mali has a population of seven millions; nearly all of them live in rural areas and 51 percent are female. The great majority, estimated at more than 98 percent, of the women are illiterate. This inability to read or write reduces considerably the contribution women can make to the economic life of the country.

In order to eliminate this great disadvantage and to promote women's literacy in particular, the Ministry of Education created a Division for the Promotion of Women (DPW) within the National Department for Functional Literacy and Applied Linguistics (DNAF-LA). The Division was in fact set up to examine the special problems of this branch of education and to devise solutions.

Its investigations revealed some common group problems of women in all parts of the country. Their daily responsibilities were demanding and time consuming. They were the water carriers and fuelwood finders; they had to work in the fields, care for the children, and prepare the family meals. They had no time to become literate. Their husbands were usually reluctant to allow them to attend literacy classes. The women themselves lacked the motivation.

The Division developed a course to help these women increase their capability to cope with the increasing demands of a modernizing society. Since 1977, a programme called *The Participation of Women in Development* has been going on. It is geared to the essential concerns of women and it aims to assure them of better living conditions.

In 1980, the Government of Mali, with financial help from the UN Fund for Population Activities and technical assistance from the UN Educational, Scientific, and Cultural Organization, set up the project called *The Use of Video in Education and Promotion of Women to Support the Literacy Courses and to Increase the Effectiveness*. It explains itself.

The primary objectives of this programme are to inform, motivate, and organize disadvantaged populations to become aware of and be able to assess their problems; to use mass communications materials to encourage them to participate in the search for solutions, to contribute to harmonious socioeconomic development.

At first, video was viewed as a sophisticated system slightly out of reach of those not experts in communications technology. In fact it soon proved itself to be an excellent tool for both informing and training. Almost anybody can learn to use the equipment and even basic repairs can be done by someone with no great technical background.

Videotape runs on roughly the same principles as sound-recording tape, with the additional dimension of vision, recording sound and taking moving pictures simultaneously. The tape is cheaper than film, does not demand processing or development, and, a matter of very great significance to countries like mine, it can be used, erased, and reused a great many times.

The operational techniques are uncomplicated and the equipment is sturdy, unaffected by climatic differences and does not normally present repair problems which a technician with some basic training cannot handle. The recordings can be played back immediately, and the playback itself can be adjusted to suit the audience.

These features allow the material to be moved directly from location to location and even used for different purposes, without having to be "treated"; there is no need for studios or laboratories or broadcasting stations. Let me quote to you some of the remarks on the medium by Martha Stuart, that dynamic American Indian who has promoted the use of video in village development in so many countries and worked closely with us in Mali.

"Video's ready ability to capture and transmit a slice of life," she wrote recently, "whether in terms of the operation of a specific programme or of a discussion among those affected by it, make it an ideal medium—with greater range, economy, and speed than film, for example—through which to involve even the most isolated villagers in development activities, as players rather than as pawns.

But there are infinitely more creative and more powerful ways to employ video at village level. The means it provides for villagers to speak for themselves, revolutionary in its own right, need not be confined to countering the prevailing information flow and bringing the villagers' voices to the official's office.

"Video's greater range of possibilities can include new horizontal channels of information flow that directly link person to person, group to group, village to village. In this sense, video technology makes entirely new and different forms of sharing possible."

It is, of course, logical for this sharing to be international; village problems and rural development are very similar in the developing world, and perhaps elsewhere too. Martha Stuart will certainly agree. She has been working for this for years and, jointly with the Tokyo-based United Nations University, has now started a project called Village Network.

In fact, my country hosted, in October 1983, the gathering of twelve national groups—from Antigua, China, Egypt, Guyana, India, Indonesia, Jamaica, Japan, Nigeria, Zimbabwe, the American Indian Nations, and of course ourselves—who are now the founding [members]. The door is open for others to come in, and The Gambia has already been invited.

The participants were all, needless to say, involved in development at the village level and interested in the medium of video. I was able to show a tape of the Malian programme and demonstrate our work through a day in a village called Sougoula.

Since we began our project, in June 1980, we have reached some 3,000 people in thirty villages in three of Mali's seven regions. We have produced programmes on women's activities, on health, nutrition, agricultural cooperatives, daycare centres, literacy, income-generating projects, and so on. All this was done with one portable production rig and one editing unit.

As I told the meeting, we have found that video is an excellent tool for sensitizing and motivating. Because the programme can be seen as soon as it is recorded, it provides real feedback; it contributes to changes in perception which lead to decisions at all levels. At the meeting in Bamako, we agreed by consensus on a sort of theoretical design for Village Network.

For a start, we decided to have a secretariat and to base it, for the first year at least, in New York, where facilities for duplication and transfer are available. We also worked out ways of building up a bank of videotapes, which of course will be in constant use in some village in one country or another. Many details remain to be finalised but some broad strategies are clear.

As a network, the targets the programmes within each country would want to reach are: the political leadership, the public service at senior level, the private sector including especially local manufacturers of consumer products, and naturally the villages and small groups who would be the ultimate beneficiaries and must be encouraged to be involved in the network.

We thought, and I say this as the consensus view of the meeting, that the network should stress the communications component in all development budgets. In the report of our meeting, you will find, apart from the Ministries concerned with foreign affairs and information, a long list of others:

> community and cooperative development; women's affairs, labour; rural development; agriculture and hydraulics—that is, water, drainage and irrigation; education and culture; energy, especially such alternative and nontraditional forms as biogas, solar, wood, gas, wind, and charcoal; forestry and reforestation; health; family planning; hygiene; nutrition; social services for the handicapped; fisheries; appropriate vessel technology; economic planning; and finance.

As you see, the list is long and practically every planner is involved. In addition, we thought it should be emphasized that video was appropriate and needed, in rural communities, in a number of uses. The report lists them as:

> training, organization and management and marketing in cooperatives; literacy, preschool, and adult education; cultural activities and entertainment; education for production, handicrafts, and farming; leadership training and self-help activities; training and sensitizing rural leaders; attitudinal training and building self-confidence; encouraging self-reliance, such as consumption of local foods; appreciating and understanding other communities; and resolving community, regional, racial, tribal, religious, and class conflicts.

Video will not do any of this miraculously on its own, of course, but it is a suitable, possibly the most suitable, instrument for many situations. This what rural development is about. In Mali, the Video Team is planning to extend activities into all seven regions. We are working at the same time to increase general interest in video.

We are cooperating or building cooperation with public and private bodies, like the Ministries of Rural Development, Agriculture,

Energy and Mines, Health, Finance, State Enterprizes, as well as with the Development Bank of Mali and the National Union of Malian Women.

We have found that some subjects, apart from agricultural development, are of special interest to our villagers. They include desertification and how to fight the process, women's activities, literacy, the physically handicapped, family planning, nutrition, health, handicrafts, and other income-generating projects.

As we extend our programme countrywide, we are determined to be self-reliant in terms of maintenance and repair. So we are going for training technicians just as fast as we are adding to our equipment.

We now have five technicians and four literacy workers who have been trained in the use of video equipment—camera work, handling of decks and television, et cetera—by video consultants. The Video Team in turn has introduced fifteen rural leaders to portable video production—camera, image sequence, programme planning, playback, and so on. Video has also assisted in training 105 rural literacy teachers.

When it comes to the network, we may have some additional problems. One which has been mentioned is that we produce in one video system, whereas another system is more commonly favoured in our region. We are also short of equipment and accessories, certainly in terms of wider participation. This means we have to upgrade our repair and maintenance capability within the country.

The network, to exchange programmes between projects, is desirable. It would contribute to the development of video as well as reinforce cooperation between countries. Within Mali itself, the numerous requests to use our equipment by other services and organizations testify to its impact.

CHAPTER 2

The Historical, Socioeconomic, and Political Implications of VCRs

We have all become so accustomed to audio- and videotape technology and its use in home entertainment and broadcasting that we often take the development of this sophisticated technology for granted. Even as late as 1960, anticipating such technology for home use at affordable prices for the United States market—much less the developing world—was beyond imagining. Some knowledge of the development gadgetry and of television broadcasting technology with regard to differing international line and color standards is helpful in understanding the way in which video spread.

EARLY RECORDING DEVELOPMENTS

There were antecedents to the audio tape recorder. For example, in 1885 C. S. Tainter, with Chichester Bell, Alexander's cousin, patented a wax cylinder alternative to his famous cousin's tinfoil phonograph. In 1888, Oberlin Smith of Cincinnati, an American engineer, published a magazine article in which he discussed something like magnetic recording wherein iron particles could be embedded in thread for later production of sound. Valdemar Poulsen, the "Danish Edison," among his many engineering accomplishments, patented in 1894 the Telegraphone, a device that stored sounds on magnetic steel cylinders or spools of piano wire. Poulsen's invention was superior in several respects, but most notably because it was entirely electromagnetic. When the Telegraphone was demonstrated during the 1900 Exposition

Universelle in Paris, it reportedly caused a sensation, and spectators lined up to hear their own voices reproduced. On November 13, 1900, Poulsen was granted patent No. 661,619 on the Telegraphone, and a few of the devices were used in the United States and Germany by law enforcement agencies and the military for the recording of messages. But the time was not right for the public to acquire the Poulsen machine; engineers and businessmen were lethargic about the device that appears to have influenced the development of the German magnetophon (Angus, 1973).

Research on something similar to audio tape recording continued in the United States and Europe in the 1920s and 1930s. One problem was an uncertain niche for such a machine in the marketplace: inexpensive female labor had created an abundance of stenographers in business, and the home market did not seem viable enough for mass production. By the 1930s, broadcasters, particularly European organizations such as the British Broadcasting Corporation (BBC), were interested in a machine that would reproduce high-quality sound and at the same time allow programmers to edit material. Such a device arrived at the BBC in the very early 1930s. The Blattnerphone, named after its German inventor, Professor Blattner, recorded sound on metal tape that could be edited by cutting with shears and then soldering. This machine remained experimental because of an annoying "plop" that was picked up as soldered pieces passed by the playback head. Also perturbing was the recorder's seeming inability to play material at the same speed at which it was recorded (Angus, 1973). Blattner's fellow countrymen at large engineering and manufacturing companies, such as AEG and BASF, had both the electrical and chemical engineering knowledge to place iron oxide on "tape" and to manufacture audio tape recording equipment. By the time this all came about, Hitler's military establishment was also interested and helped push the development of the recording technique for its defense application.

John Mullin, who would become involved after World War II as an engineer with California-based Bing Crosby Enterprises, remembers being fascinated with the quality of sound he heard while he was an officer stationed in Britain during the war. One evening, he recalls listening to German radio after BBC had signed off and being impressed with the technical quality of Strauss and Lehar music and Viennese operetta arias. "The sound was so flawless that we were convinced we were hearing live performances. The usual deficiencies of record scratch and other telltale distortions were completely absent" (Mullin, 1979, p. 80). Later, Mullin would see tape-recording machines left in Paris by the departing Germans, and he would hear the magnetophon demonstrated in occupied Germany.

POST–WORLD WAR II TECHNOLOGY

While Germans had invented the first operational audio tape recorder (the magnetophon) and the correspondingly important tape, Americans brought tape recording into popular use in broadcasting. John Mullin and J. Herbert Orr (creator of Irish brand recording tape) are credited with bringing the first magnetophons to the United States. It was on these modified German-built machines that Mullin demonstrated his faith in magnetic tape recording to Bing Crosby and his producers. He had been invited to show the advantages of tape to Crosby, who—unlike virtually everyone else in all-live network radio—preferred to record his shows (Mullin, 1979). Crosby was suitably impressed and the result was threefold. First, in August 1947, Crosby's producers used the liberated magnetophons to record and edit a show to be broadcast on October 1. But that actual October 1 broadcast originated from disks; network officials, still skeptical, had the program's tape placed on disk for transmission. It was not until May 1948 that the Crosby show went out directly from tape (Angus, 1973). Second, Crosby created Bing Crosby Enterprises to continue developing tape technology. This venture was later sold to Minnesota Mining and Manufacturing (3M) when Ampex developed a higher-quality recording machine (Ziff, 1979). Third is the creation of Ampex itself, an organization synonymous with broadcast-quality, audio tape technology, but a company probably better known for its contributions to video tape technology.

THE DEVELOPMENT OF VIDEOTAPE

In 1986, United States television broadcasters marked the thirtieth anniversary of video tape recording. It was in May 1956, at the National Association of Broadcasters covention in Chicago, that the Ampex Corporation officially demonstrated its broadcast-quality video tape recorder. Ampex had scored a breakthrough in development by solving a problem plaguing others who were working on videotape by innovating the four rotating recording head system. This development, known as quadruplex recording, or "quad," allowed two-inch tape to travel at less than super speeds and the requisite complicated video information to be placed on it. Of course, by today's standards, these early models were crude: there were problems with "banding," caused by head alignment problems; and mechanical editing was all but impossible. Electronic editing had not been invented.

Although the three networks were skeptical of videotape, they were not as hesitant to experiment as they had been with audiotape nine years earlier in radio. The cost advantage of tape for delaying broadcasts for other time zones was obvious. CBS was the first to air a program on videotape, when "Douglas Edwards and the News" was telecast on November 30, 1956, from CBS Television City in Hollywood—a three-hour delayed broadcast of the evening news from New York. In the following months, the other two networks followed CBS's example (Anderson, 1979).

The quadruplex video recorders were large and heavy, even the solid-state advanced models that appeared in the 1970s. When "portable" units were developed, they had to be put in remote vans. In the United States, Europe, and Japan, the development of helical-scan, or "slant-track" recording, moved forward. Machines using this technology were smaller, but the main problem was quality, particularly that demanded by television broadcasters. Meanwhile, the Japanese introduced portable reel-to-reel machines for the educational and industrial market. Sony was successful with its half-inch tape on a seven-inch reel CV and later AV series in the United States and elsewhere. International standards for helical technology were finalized, and in 1969 Sony first introduced the three-quarter-inch U-Matic recorder—a machine that transcended broadcast studio, industrial, and home use.

By 1975, when Sony started selling the Betamax home video recorder, virtually all the conditions were right for the new technology. First, electronic circuitry was advanced enough to be mass produced for the home market. Second, the psychology of the VCR cassette itself—a tape encased in plastic the size of a paperback book that freed nontechnically oriented people from reel-to-reel tape phobia—placed video operation within reach of the home consumer. Third, it appeared that television viewers, to judge from the enthusiasm of customers for the Betamax and later VHS units, were ready for an appendage to their television sets, one capable of time shifting and playing prerecorded material.

THE CULTURAL IMPACT OF VIDEOCASSETTES

The scattered literature on the cultural impact of videocassettes makes for fascinating, but schizophrenic, reading. Much of it is anecdotal or theoretical, with very few systematic studies. Nevertheless, although film and television managers (businessmen that they are) concentrate on the economic significance of video's intrusion in the media realm, some governmental, religious, and educational figures (and of course, parents) have concerned themselves with the possible impact of video upon morals, values, behaviors, and other societal traits.

One key impact of video has been postulated in the areas of entertainment and use of leisure time. Because of its relatively low cost and in-home use, video has both provided entertainment where not much existed before (such as the Gulf states, or small towns and semiurban regions of the Third World) and substituted for regular television and cinema viewing. Video has been targeted as having aggravated the sedentary nature of what Gubern (1985) terms *homo electronicus* (electronic man) and *homo otiosus* (leisure man). According to him, people have found in the VCR a suitable instrument for their increasing leisure time, because it offers diversification of content, decentralization of source, feedback, privacy, and mastery of programming. Obversely, the VCR has been blamed for hindering the work ethic. Some examples have been given where watching video has affected production; in Taiwan, late viewing of video has meant less labor effectiveness in the morning hours, and in Indonesia, fishermen have not wanted to work at night because they would miss prime-time television and also video.

The potential impact of the VCR upon education has divided critics, with some claiming that video, by providing an alternative to reading, has adversely affected literacy, while others point to programs utilizing video that enhanced literacy and other educational goals. Numerous examples abound; in Thailand, farmers teach each other with the use of video; in Ghana, the VCR was implemented to increase the effectiveness of management trainers; throughout Latin America, children are taught sexual responsibilty through music videos featuring two popular singers, Johnny and Tatiana. In Nepal, Kenya, Mali, and India, among other countries, women have been educated, and their conditions generally uplifted, through video training schemes, video bulletin boards, and video letters.

Another area where there is a diversity of viewpoints on the impact of video is family life. There are a few like Gubern (1985), who associated video with positive values upon family life, claiming that in a fragmented society, the VCR can foster family cohesion. But he says elsewhere that the VCR has substituted the superficial smile of public relations, as seen on video, for genuine human contact. A 1981–1982 study in Sweden (Baboulin et al., 1983) disputes the family cohesion concept; it found that children do not watch video with their parents (only 2 percent do) but, instead, with their friends (77 percent). Baboulin and associates (1983) argue that the VCR actually coincides with the changing pattern of family life, what they call "selfishness without regrets" or "to live separately together." In the Third World, where the extended family is still prevalent, video may have a cohesive effect in pooling economic resources for financing the hardware, but an adverse effect upon family conversation and other social intercourse, keeping family members glued to the screen.

The portrayal of explicit sex, violence, and pornography on videocassettes has been cited as having a negative effect, especially upon children, as well as more generally upon religious and traditional values and morals. The few studies concerning children and videocassettes have been carried out in the Western world. In England, Barlow and Hill (1985) said there was conclusive evidence of causal links between violent behavior and the viewing of violent videos, while others stated the causal link is still unproven. Roe (1983), studying Swedish adolescents, said their watching of video with friends is a signal of distancing from the adult-dominated culture, which labelled them as failures. Although not based on conclusive research, 1984 police reports in Bangladesh linked VCR usage with the sharp upsurge in crimes of rape, murder, robbery, hijacking, and kidnapping.

Religious authorities, especially in Muslim countries, nearly all of which are in the Third World, have been worried about video content; not only about pornography, obscenity, and excessive violence, but also about themes that depict disrespect for authority, a more liberated role for women, or greedy consumerism and commercialism—all anathema to certain religious and cultural norms. As governments of Malaysia, India, Indonesia, Nicaragua, and Peru, among others, have tried to stop television that promotes consumerism (Boyd & Straubhaar, 1985), it is only natural that they would especially want to halt video that allows even more freedom. The passivity, consumerism, and materialism that scholars over the years have attributed as cultural blemishes of television are even more readily available to video viewers in the Third World.

Newly emergent countries, which encompass most of the Third World, have set up numerous national development programs that they feel will be compromised by a medium such as video. In Malaysia, a former information minister castigated video because of its tendency to take audiences away from government-sponsored television promoting developmental messages. A large part of that audience was composed of Chinese who wished to see films and television shows on video, in their ethnic language, imported from Hong Kong or Taiwan. In multiethnic and multilingual societies, such as Malaysia, video can keep alive original languages to the chagrin of authorities who are hoping for the quick assimilation of diverse populations. Similar patterns have been perceived among Third World immigrants in Europe, the United States, and elsewhere, where video has been used to keep them in touch with their original homelands, much in the manner as Frantz Fanon said happened during the age of imperialism, when colonialists developed and sought media to keep them in touch with "civilization, their civilization."

Because VCRs and cassettes travel so rapidly and relatively inconspicuously, they penetrate even the most closed societies, whether they be Muslim as in Saudi Arabia, or communist as in China. The result is that national and ideological policies imbued in the people for years without much outside interference now must cope with polished, contrary views brought into homes by video.

Other writers have noted the capability of VCRs to cause economic imbalances, to widen gaps between the haves and have-nots and to raise frustration levels in the Third World. In Africa, the high cost of VCRs—twenty-nine times the minimum salary in twenty-nine nations—keeps them the exclusive privilege of socioeconomic elites and, according to one African, "reinforces the potential for cultural alienation and conflict" (Jouhy, 1985, p. 434). Jouhy, writing about the Third World, said television and video viewing induce ever-increasing social and psychological imbalances, because the compensatory escape fails to provide an occasion for participating and acting out people's needs, thus adding to the feeling of frustration and craving. He added:

> With each television set or video recorder that we export via the one-way path to the Third World, with each show or cassette that we sell in these parts, we transport the germ of unrest, of instability, revolt, delinquency, and violence, because the practical means of attaining the fictive goals that the media delivers [sic] are missing (p. 436).

On the other hand, Jouhy seemed as uncertain about the impact of the VCR as the next person, attributing to it the potential for collective creativity among the poor and oppressed. For one thing, television and VCRs show normally fatalistic peoples that the general conditions of human beings and society are not shaped in heaven (that the misery of their lives is not because of fate), but rather on earth. Once the people realize that, according to Jouhy (p. 436), they will ask why they are being exploited and how they can change their lives. The new media, he contends, haul "the spectator out of the isolation and hopelessness which is inherent in the contradiction in which he lives between reality and manipulated dreams" (p. 437).

GOVERNMENTAL POLICIES ON VIDEO

Because of these potential cultural effects, most Third World governments have been quick to denounce video as a menace. When coupled with political uses not sanctioned by—in fact, contrary to—the authorities, video becomes an instrument to be feared. Among political uses made of video are circumventing controlled media in political crises, propagandizing guerrilla activities to peasants, training antigovernment forces, providing inspiration and guidance to would-be assassins, frustrating government efforts at television censorship, and releasing news for lobbying purposes (Ganley & Ganley, 1986).

Third World governments have deliberated long and hard on the control of video; they have established monitoring and regulatory bodies, and pleaded with the public to respect copyright and other media-oriented laws. Most of what they have attempted has not been fruitful.

A number of factors account for these failures. Prominent among them is the elusiveness of video, which is independent of a national media system that can be controlled and is brought into the home by individuals, not signals. The independence of video allows viewers to fulfill their wildest fantasies, including, according to the Ganleys (p. 6), "the wildest fantasy of all, [that] VCRs and videocassettes symbolize control over Hollywood and other such dream empires."

The international black market also prevents effective control of video. In numerous Third World countries, most (or even all) VCRs and videocassettes are smuggled in. Vehicles used are well-established smugglers' rings, like, for example, the one operating between Dubai and Bombay, as well as tourists' and migrant workers' luggage and diplomatic pouches. The profits are rewarding enough in countries such as India that it pays black marketeers to hire video carriers and pay their air transport to other countries to bring back illegal VCRs and cassettes.

Connected with the smuggling is the bane of governments and film and television industries—pirating. As early as 1983, the world video piracy take was estimated to be over U.S.$1 billion; in the Philippines, practically all (99 percent) video was in pirate hands. Despite all the hoopla about piracy since 1983, the overall situation has not changed markedly. The pirating of cassettes occurs in various ingenious manners. In some instances, pirates record television shows or films in one country and send the master abroad for mass duplication. For example, Japanese television shows have been taped off the air in Tokyo, and the tape given to a traveler on a late-night or early-morning flight to Taipei, where it is made into multiple copies for sale the following day. Much pirating is also carried out within a given country, either by duplicating prerecorded cassettes or by taping directly off the screen. Film directors in the Philippines and Taiwan said pirates bring cameras into theaters and record films live; the heads of theater patrons can often be seen on the cassette and the laughter of the audience can be heard. In still other cases, unethical film producers and processors make illegal copies for sale to pirates. This has happened in India, where master prints sent to labs for processing were pirated.

Concerning policing, the governments themselves are caught in no-win situations. The economic viability of many depends upon advanced black market systems, a fact that keeps them from controlling too strictly. Furthermore, conflicting desires by governments to placate and balance conservative and liberal forces have prevented effective control of video. Other governments ignore illegal video because it takes the pressure off them to improve national broadcasting, which they may not have the resources to do. Still others are either so concerned with survival that they cannot think of videocassette control as a priority, or they are baffled by a problem that descended upon them so suddenly and with such force. In short, most Third World governments have been mild in their frustration; and when they have clamped down, most times it has been for economic, not political or cultural, reasons. The Ganleys (1986) suggest that governments take some joy in having short-circuited via piracy and smuggling the long wait for United States films.

Whatever the reasons for cracking down, Third World governments have attempted a variety of control mechanisms, most to no avail. Probably the most extreme are imposing the death sentence for illegal trafficking in video and the outright banning of VCRs and cassettes. Iran has tried the former, but illegal video still persists. From 1981 to 1986, Mexico banned VCRs and cassettes, after which it put their control in the hands of a television conglomerate, Televisa's Video Visa. Another control route taken is that of nationalization. But it too can be manipulated. When Indonesia nationalized three companies to handle all video in the country, door-to-door video rental salesmen peddled poor-quality, other-than-first-generation copies. One report stated that 90 percent of the videos continued to be illegally imported.

India and other countries, for a short time, regulated the acquisition of, and access to, VCRs and videocassettes by producing their own. But the Indian video industry, unable to cope with imported competition, virtually died before it got off the ground. Even at 190 percent duty, imported VCRs could be sold in India for a price less than two-thirds that of local brands.

In many countries stiffer penalties for piracy and copyright infringement have been imposed, and video has been included in updated copyright legislation. Again, much of this has backfired. For example, the Singapore government early on ordered that all cassettes offered for sale carry a censor's certificate, which prompted some critics to accuse the authorities of encouraging piracy. They based their accusation on the fact that no questions were asked about copyright ownership when application for a certificate was made. Pakistan, among others, has required the licensing of VCRs; but the video phenomenon persists and all types of cassettes, including pornography with Pakistani participants, are available under the counter.

Authorities emphasize that because of the privacy of home video, the public's reluctance to cooperate with the authorities, the large number of rental outlets, and the relatively few policing agents, legislation is difficult to enforce. Some governments have tried to educate the public to avoid illegal video, but the economic reality of less expensive—if poorer quality—cassettes looms larger than the ideal of civic consciousness.

As indicated earlier, the surprising speed with which video descended on the Third World caught governments off-guard. The copyright legislation in countries that subscribed to international agreements usually was embarrassingly outdated, limited to cinema and sometimes going back to colonial days. Thus, many countries, such as the Philippines, have only recently gotten around to amending the laws to include video.

Even with the few success stories of the abatement of piracy, other factors have been credited. For example, the government of Hong Kong in recent years assigned more than forty agents to control video piracy, with seemingly effective results. However, film and television managers in the British colony give another reason for the comparatively low use of video; they claim it results from crowded, undesirable housing and the desire of residents to enjoy the many exciting attractions on the streets.

In sum, although Third World governments are aware of the need to regulate video because of the threat it poses to national development and cultural programs, to local television and film industries, and to revenue-generating taxation, they are stymied by the problem and realize that the technology has run far ahead of legal and scientific preventive barriers.

THE ECONOMICS AND ECONOMIC
IMPLICATIONS OF VCRS

The economic situation of Third World countries has a great deal to do with whether VCRs spread widely within their borders and with how they are used. To start with, affecting VCR diffusion are the cost of VCRs, the ability of various nationalities to afford them, the varying ability of different classes of people within countries to buy VCRs, and the role of some institutions in buying and spreading VCRs.

The cost of VCRs varies widely for a number of reasons. The world price of VCRs has dropped rapidly over the last few years, from an average of about U.S.$500–$600 in 1983, to an average of U.S.$400 in 1984, to U.S.$200–$350 in 1988, This decline in prices has made VCR diffusion an equally volatile issue. Overall, the plummeting prices have added a tremendous incentive to the spread of VCRs in the Third World.

Before the price drop, the new medium was largely inaccessible in many countries for purely economic reasons. Out of 126 countries for which data were reported by the World Bank (1986), the average income per year was under U.S.$1,000 in fifty-six of them. Another thirty range from U.S.$2,000 to $3,000. The area of economics is one in which the Third World is perhaps most divided. Within the Third World, one must cover the countries mentioned above as well as newly industrializing Singapore and Hong Kong and oil producers such as Kuwait and Saudi Arabia. It is well to remember, particularly in the Third World, that average income figures, such as those available from the World Bank, often mask a very skewed or unequal distribution of wealth. In a few countries, such as Brazil, about 50 percent of the income is held by less than 10 percent of the population, while the comparable figures for the United States and most Western European countries range from 20 to 25 percent (World Bank, 1986). In situations of extreme income concentration, it is difficult for the poorer part of the population to acquire a radio, much less a VCR. What is surprising in many countries is the number of relatively poor people who sacrifice other interests to acquire first a television and then a VCR.

Prices are remarkably varied, however, largely because of government decisions to place high tariffs or taxes on imported VCRs. Most Third World governments consider VCRs to be luxuries; and while many countries do not bother to collect taxes on such imports, many others do. Tariffs can be as much as 300 to 400 percent on VCRs in some Third World countries. Even more restrictive, some countries, such as Brazil and Mexico between 1982 and 1986, have simply banned the importation of VCRs to reduce balance of payments problems and cut spending on what are considered frivolous luxuries. While some countries restrict the entry of VCRs for political reasons, fearing their impact as an uncontrolled medium, purely economic policy reasons lie behind many, if not most, of the national restrictions (both import barriers and tariffs) surveyed for one recent study (Straubhaar & Lin, in press).

Tariffs and import barriers lead to smuggling and the sale of VCRs on black markets. In many cases, over 80 percent of VCRs in a country have been smuggled in (Ganley & Ganley, 1986). Thus, VCRs fall into the illegal sector of many Third World countries. In many countries, legal channels do not meet consumer demands for a number of reasons: bureaucratic controls; inflation; price controls; and high value-added, sales, or luxury taxes. In such economies, it is very difficult for anyone, including the VCR manufacturers and the government, to know how many VCRs have been brought in, except by conducting market surveys. Estimates of the number of VCRs in various countries made by the Motion Picture Export Association of America and others often vary by hundreds of thousands, particularly in some of the larger countries, such as Brazil.

Because of VCRs' increasingly proven attraction to consumers, several Third World countries are trying to build VCRs at home. These nations include South Korea, which has begun to cut into Japan's export monopoly on the machines, Argentina, and Brazil. Countries that do not have the manufacturing base for local companies to build VCRs have allowed multinational manufacturers to build assembly plants to service the local market and engage in some exporting of the machines to other markets. Such operations are particularly common in duty-free ports, where import duties are waived or reduced (Mattelart & Smucler, 1985).

Videotapes are subject to similar constraints. They are often subject to heavy tariffs and controls, both for blank and prerecorded tapes. Blank tapes cost anywhere from U.S.$4–$5 to over U.S.$50, depending on national tariffs and controls. Prerecorded tapes vary even more in price, particularly because of widespread smuggling.

The cost of prerecorded tapes in the United States and other centers of production has been dropping rapidly in the last year or two. Before 1985, however, the high cost, or simple unavailability, of legitimate prerecorded tapes created a situation in which piracy prevailed. Throughout Latin America, for instance, 80 to 100 percent of the rental or purchase tapes available are pirated (Ganley & Ganley, 1986).

The incentives for videotape piracy are largely economic. Obviously pirate copies can be produced more cheaply, since no royalties are paid. In Sri Lanka, for example, music tapes of local artists are more expensive than those of foreign artists, since tape copiers pay royalties only to the local artists (International Association for Mass Communication Research, 1985).

Furthermore, piracy has been encouraged by a combination of economic and legal factors when national tariff, copyright, and censorship barriers have held back distribution of legitimate copies. In many countries, a final factor was the practice of film distributors not to bring films into theatrical release, much less video release, until after pirated copies were already available. Initially, distributors held back videotape release to maximize profits in theatrical release; but, as in Mexico and Venezuela, distributors are increasingly finding it economically necessary to advance release times of films on videocassettes in order not to lose the entire video distribution market to pirates.

Most pirated videotapes are imported entertainment, either feature films or series. That in itself is partially due to economic factors underlying previous patterns of television and film importation; in many countries, importation of such films and series is much cheaper than producing them locally (Lee, 1980). Still national films are very popular in Brazil, Egypt, India, and other film-producing nations, where industries have been built up to support film production for the local market. These industries are now threatened by the fact that piracy cuts into their financial support through theatrical release or sale to television.

If VCR use for watching pirated videotapes is to some degree determined by economic issues, so are some other aspects of VCR use. The viewing of pirated tapes falls within the dominant use of VCRs— i.e., watching entertainment. However, other forms of programming can be used on VCRs. Evidence from Brazil, India, and Nigeria indicates that as the cost of video production equipment drops, then a variety of local enterprises spring up to videotape weddings, funerals, births, family gatherings, and religious ceremonies. Video production also increases by religious groups, unions, political parties, or community associations to spread their point of view ("Videocassete no Brasil," 1986).

Finally, it must be remembered that the real economic impact of the VCR on industrialized countries is yet to be fully investigated. However, preliminary data strongly suggest that the way Hollywood markets its films has been permanently altered; and, at least in the United States and some European countries, television viewing patterns have changed. In these pages we are primarily concerned about the Third World, those countries generally classified as developing.

PART 2
Regional Perspectives

CHAPTER 3

Videocassette Recorders in the Arab World

As a true revolution, the emergence of video has been like an explosion, full of energy and even passion, provoking divisions, inflicting
wounds, destroying, upsetting and, also, raising hopes and promis-
ing new riches in the so-called "television wasteland" ("Video: A
media revolution?," 1985, p. 1).

The terms Arab world, Middle East, and Near East are often used
interchangeably, but incorrectly. The designation Middle East is British
in origin (Koppes, 1976), reflecting the considerable influence over the
vast area that the island nation had at one time. We have chosen to use
Arab world because it reflects more accurately the geographical area of
interest. This chapter does not include Israel, Turkey, Iran, Cyprus, or
Afghanistan, obviously not Arabic-speaking countries.

The Arab states in this study—Egypt, the Sudan, Lebanon, Syria,
Jordan, North Yemen (Yemen Arab Republic), South Yemen (People's
Democratic Republic), Iraq, Kuwait, Saudi Arabia, Bahrain, Qatar, the
United Arab Emirates (U.A.E.), Oman, Algeria, Libya, Morocco, and
Tunisia—are members of the Arab League, an association of Arab
countries founded in Egypt after World War II. League headquarters
were relocated in Tunisia after Egypt signed a peace treaty with Israel
in 1979. Although Somalia and Mauritania are League members, they
are not included here because they are not historically Arab. The
Palestine Liberation Organization (PLO) is a member too, but Palestine
no longer exists as a state; Palestinians do operate a government in
exile.

The Maghreb, i.e., North African Arab countries west of Egypt,
are not a focus of this section, and some states—most notably the
Arabian Gulf[1] countries (Kuwait, Saudi Arabia, Bahrain, Qatar, the

U.A.E., and Oman), Jordan, and Egypt—are dealt with in great detail, because more is known about video there and because they play an important role in both videocassette recorder[2] (VCR) hardware and software distribution. As will be seen in this chapter, the Gulf states play a unique part in Arab world VCR penetration.

Even the casual newspaper reader or broadcast news listener/ viewer knows that politically the Arab world is not always a happy place. Within this decade there has been conflict between both Egypt and the Sudan and Libya. There is the continuing Gulf war, pitting Iraq, supported with various degrees of enthusiasm by some other Arab countries, against Iran, supported by Syria, Iraq's archrival. Lebanon continues to be torn by both internal and external interests.

This region contains its haves and have-nots; in economic terms, there exists an uneven distribution of wealth. However, the gap between rich and poor grew wider after the 1973 Middle East War when the Gulf states, following the example of King Faisal of Saudi Arabia, first banned the export of oil to the West and then resumed shipments at quadruple the 1973 prices. This act enriched an already prosperous area of the world—the Gulf states—and indirectly added a measure of financial well-being to the less fortunate countries in the region. The Gulf oil boom needed manpower to accomplish the area's ambitious development plans, and expatriate workers came from other Arab states, the West, and Asia to toil for attractive wages. In fact, an understanding of Arab world oil economics is important to the comprehension of the Byzantine complexities of Arab electronic media; such an understanding is absolutely essential to knowing the magnitude of the region's VCR phenomenon.

Arab countries talk a great deal about unity, stating that they are essentially one nation sharing a common language (Arabic), a history, and a religion (Islam). It is true that the vast majority of Arabs are either Sunni or Shiite Muslims, but there are Christian minorities in several countries, most notably in Egypt, the Sudan, Jordan, and Lebanon, the country with the largest Christian minority[3] in the Arab world. There are, of course, common historical experiences in the region, but few Arab countries have similar recent histories. Politically, Arab nations differ from each other. All Gulf states, Jordan, and Morocco are headed by monarchs; some states elect their officials. Politically, Kuwait, for example, with a head of state determined by the royal family and an elected Consultative Council (Parliament), differs dramatically from Marxist South Yemen or Islamic-socialist Libya.

The Arabic language allows individuals from Morocco to Iraq, and Yemen to Lebanon, to communicate with one another. While spoken dialects are very different, nevertheless a person from northern Iraq can talk with an Algerian by using classical Arabic, the language of the Koran. Since the 1930s, yet a third form of Arabic—neoclassical, or modern standard, Arabic, the language of newspapers and the broadcast media—has gained widespread use. The Egyptian film industry, active since the late 1920s, made the Cairene dialect widely understood, even before television arrived in the Arab states.

At least historically, politically, and culturally, the Arab countries do not constitute a homogeneous region. Yet it is relatively easy to discuss the electronic media in common terms, because Arab radio and television are similar in organization, philosophy, and operation. There are several factors important to understanding the situation regarding Arab television in particular.

First, there is international television viewing in some parts of the Arab world. During hot, humid weather, television and FM radio signals "tunnel"—i.e., travel beyond the line of sight—in the Gulf, Red Sea, and Suez Canal areas. A viewer in the Eastern Province of Saudi Arabia, using a roof-mounted antenna and rotator, may see as many as thirteen different channels, ten of which are non-Saudi signals—two each from Kuwait, Bahrain, and Qatar, three from the U.A.E., and one from Oman.

Second, in Arab states, all official television is either operated directly by the government or through a quasi-government corporation, usually one that either is part of, or reports to, the Ministry of Information (MOI). Since the start of the Lebanese 1975–1976 Civil War, several pirate radio and television stations have begun. Indeed an oddity is Middle East Television, a station operated by evangelist Pat Robertson's Virginia-based Christian Broadcasting Network (CBN) in South Lebanon, with Israeli support.

Third, Arab television broadcasters get little or no formal scientific feedback from the audience. It is not the practice of these broadcasters to undertake either studies providing information about program preferences or research exploring the uses of television among viewers. Most reliable data come from surveys commissioned by manufacturers wanting information about consumer brand preferences or from the British Broadcasting Corporation (BBC), the Voice of America (VOA), or France's Radio Monte Carlo Middle East (RMCME), all Western international radio broadcasters gathering audience estimates.

Fourth, the majority of Arab television systems permit advertising. However, commercial income provides only a small portion of the systems' operating and capital revenue. Governments provide most of the financing for Arab world television broadcasting.

Fifth, some Arab states have second television channels, many of which cater to Western expatriates or to the Western tastes of their own citizens. There is a great deal of television programming imported from the United States and Great Britain on these second services.

Sixth, Arabs are enthusiastic television viewers. During the hottest months, people are more homebound and television becomes an increasingly important leisure activity. The medium is ideal for Arab culture, with its strong family ties and tradition of entertaining family and friends at home. There is little competition for television in some countries where there are no nightclubs, few cinemas, and little live theater. Public cinemas are still not permitted in Saudi Arabia.

Seventh, all Arab countries use the European-standard 625-line transmission standard, even though they have not adopted the same color system. The Gulf states, except Saudi Arabia (a SECAM, Sequential and Memory, country), use the German PAL (Phase Alternating Line), while other countries, such as Egypt and Syria, transmit color via the French SECAM system.

Finally, some Arabian Gulf states have extremely high VCR penetration rates—the highest in the world, in fact.

THE HISTORY AND USES OF BROADCASTING

Of course, the VCR is inexorably linked with television and the visual medium's antecedent, radio. Like all forms of electronic communication, radio receivers and transmitters were first introduced by foreigners, specifically Europeans. Italy's Radio Bari was the first international service to broadcast to the Middle East in Arabic. Starting in 1934 (Radiotelevisione Italiana, personal communication, Rome, October 19, 1979), Radio Bari began transmitting political propaganda to North Africa with the aim of gaining support for Mussolini's regional policies. There is no evidence that the radio broadcasts helped the italian government, but they did, in part, inspire the BBC's first foreign-language service; on January 3, 1938, the British began shortwave Arabic transmissions ("Arabic broadcasts from London," 1938).

Until after World War II, the majority of radio programming available in the Arab world came by way of shortwave broadcasts from Europe. Some transmissions by Europeans originated in Arab nations. The Egyptians started a short-lived free-enterprise radio experiment in the 1920s, but the government soon tired of the disorganized state of affairs and shut down the private stations. In 1934, the Egyptian government sanctioned state-organized broadcasting, but the effort was not indigenous; the station was built and operated by British

Marconi ("History of the U.A.R. radio," 1970). In 1937, the French started radio in Lebanon (UNESCO, 1949); and in 1936, the British government began radio in Palestine (*Palestine Department of Posts and Telegraphs Annual Report*, 1937).

The first government-operated television station in the Arab world started in Iraq in 1956, after the government purchased a British Pye television facility that had been brought in by a British company for a trade fair. However, before this, the United States Air Force had been transmitting to those living on or near Wheelus Air Force Base in Libya. Another United States Air Force station, AJL-TV, went on the air in June 1955, from the Strategic Air Command base in Dhahran, Saudi Arabia (*USAFE television story*, 1955). RCA constructed the first phase of the Egyptian television system; the United States Army Corps of Engineers, under agreement with the Saudi Arabian government, built the initial television stations in the kingdom. (For more details, see Boyd, 1982.)

Broadcasting and VCRs seem ideally suited to the Arab world because Arab culture fosters extremely strong family ties; entertainment is group oriented; electronic gadgets are status symbols; leisure time is abundant; and media receiver hardware is shared.

First, Arab family ties are much stronger than those in the West or, for that matter, in some other developing areas of the world. Even in the contemporary Middle East, married children sometimes live with the husband's parents until they are sufficiently well-off to afford a place of their own. The degree to which Arabs seek entertainment outside the house varies: the financially better-off in Egypt, Jordan, Lebanon, and Jordan occasionally visit restaurants, cinemas, and nightclubs, but most entertaining is home based and most often involves extended family members. As was mentioned before, Saudi Arabia still does not permit public cinemas or nightclubs.

Second, the image of Arab men sitting in coffeehouses playing cards, smoking water pipes, talking, and listening to radio or television broadcasts is still a reality in some parts of the region. However, affluence, particularly in the Gulf, has shifted the site of such activity for the better-off to first-class Marriott, Inter-Continental, Hilton, Sheraton, and Holiday Inn hotels.

Rolo (1941) describes a scene from the 1930s which, even in the late 1980s, is still appropriate in remote areas of Yemen, the Sudan, Egypt, and portions of the Levant:

> When the day's work was done both the *fellaheen* [peasants] and the city dwellers would betake themselves to their favorite cafés, huddle together under a fuming oil lamp, and stolidly smoking their water

pipes play game after game of backgammon until the communal
loudspeaker gave forth the voice of the Bari announcer (pp. 45–46).

Coffeeshop customers today listen to the BBC, the VOA, Radio Monte
Carlo Middle East, or Egypt's Radio Cairo or the Voice of the Arabs.
They also watch television and what is offered on VCRs. The area has
not, however, produced a large number of Indian-type "video parlors," public establishments where VCR material may be viewed for a
fee.

Third, particularly among the financially better-off, the acquisition
of gadgets, whether the latest car or consumer electronic device, is a
status symbol. This is the case in the Arabian Gulf and in Jordan,
which has benefitted from the influx of Gulf expatriate salaries. Having
a late-model stereo system, television set, video game, and a VCR
takes precedence over acquiring other home appliances such as a
washing machine. The electronic items both contribute to family home
entertainment and are visible; they are evidence of economic achieve-
ment for visitors to admire.

Fourth, the Arab world workday is short (usually from 8:30 A.M.
until 1:30 or 2 P.M. six days per week), thus leaving abundant leisure
time. The excessively warm, humid weather keeps people inside (in an
air-conditioned environment for those who can afford it) for a large
part of the year.

Finally, sharing media hardware—inviting people to listen to a
radio or stereo system or to view television—is really an extension of
Arab hospitality. The owner of a small grocery or clothing store often
uses music to attract customers. Anyone who operates a radio or
audiocassette player in public is inviting others to enjoy the entertainment.

Shortly after Oman started television, the government placed sets
in outdoor public areas so that people who could not afford receivers
could view the nightly offerings. Some sets were installed near busy
intersections in Muscat, the Omani capital, but later had to be removed
because of traffic congestion caused by viewers, some of whom
stopped their cars to see programs. A personal experience can be given
as an example of the hospitality. In 1977, during an evening walk with
an American friend through a residential area of Khartoum, Sudan, we
stopped briefly at the open gate of a house with an operating television
set on the porch. The residents insisted we join the family for refresh-
ments and the evening film on television.

The predisposition to share electronic home entertainment equipment helped the initial spread of VCRs. Tapes acquired from trips
abroad were passed among family members and friends, with copies
being made at some points along their way back to the owners.

CONDITIONS IN THE ARAB WORLD
THAT INFLUENCE VCR OWNERSHIP

Having discussed some of the characteristics of broadcasting, particularly television, unique to the Arab states, let us present five conditions
related to governmental media policy and the prevailing culture that
have promoted the rapid spread of the machines.

Limited television programming. Where only a few channels are available for certain periods of the day, people are more predisposed to
purchase VCRs in order to satisfy their own viewing tastes. The Arab
television broadcasting day is still relatively short, although as more
programming of a desirable nature becomes available, and new studio
facilities are completed, transmissions are being lengthened. There is
some morning educational television—often instructional material
directly tied to the national school syllabus—and material designed for
women. However, the norm is for transmissions on the main Arabic
channel to start during midafternoon and continue until midnight.
Second channels, mostly featuring foreign programming, start later in
the afternoon or in the early evening and stop at 11:30 P.M. On Friday,
the Muslim day of rest, television broadcasting begins in the morning
and continues all day.

Controlled television news and entertainment. Where governments, private broadcasters, or public corporations present heavily controlled
news and entertainment, viewers acquire VCRs in an attempt to gain
control of what they see. All television news (with the exception of that
on the illegal stations operated by religious factions in Lebanon) is
strictly controlled to ensure that governmental policies are followed.
Entertainment programming is closely censored in order not to violate
local cultural, religious, or political precepts. The television censorship
policy of Saudi Arabia is admittedly the most stringent in the Arab
world, and it combines all the proscriptions of other Gulf states.

Affluence. Countries with high per capita GNPs have high rates of
VCR ownership. As one might expect, Kuwait, with a 1982 GNP of
U.S.$19,610, has a much higher rate of VCR penetration than does
Egypt, with a 1982 GNP of only U.S.$670 (World Bank, 1985).

Expatriate workers. Large numbers of Middle Eastern and Asian expatriates working in the oil-rich Arab Gulf states take VCRs home
during visits or when completing assignments. Most Gulf countries do
not tax consumer electronic goods, thereby making them much less

expensive than in the workers' home countries. The Gulf states, with the exception of Oman and Bahrain, abolished all taxes in the mid-1970s, when oil income was sufficient to finance every government need.

Suppression of political activity.   In countries where political expression is limited, VCRs can become a means of communicating political points of view; but in the Arab countries, revolutionary material on video-cassettes is not yet a problem. Even so, a few plays with political overtones have been banned in the Gulf.

While these are by no means the only conditions affecting the acquisition of home videocassette recorders, they provide the basis for a discussion of the impact this consumer device is having on the Arab states.

ENTERTAINMENT PIRACY IN THE ARAB WORLD

Here again, it is difficult to discuss all Arab states collectively. However, in general, the developing world does not pay the same attention to copyright and patent laws as does the industrialized world. In part, this is because either there is no local legal recognition or enforcement of such laws, or there is a segment of the population that will not obey them. Engaging in what is an illegal activity in the West is a matter of economic survival in some developing countries. The copying of brand-name designer clothes, watches, luggage, auto parts, cigarettes, books, films, and videotapes in the Third World has become rampant despite serious efforts by some governments and manufacturers to stop it.

Another factor contributes to the unauthorized copying of software in the Arab world: the scarcity of local or legally imported material. For example, it was not until the 1950s that Saudi Arabia permitted music, with the exception of military marches, to be played on its government radio stations; and even after the ban was lifted, the musical offerings were very limited. Admittedly an extreme example, Saudi Arabia reflects the Gulf's conservative religious orientation, which often conflicts with the Western trappings of modernization. Although rich in oral history, the Gulf states have only a recent history of painting, sculpture, and drama. Islam traditionally forbids reproduction of the human form. Also, the continuous wanderings of Bedouin tribes discouraged artistic development. Thus the lack of indigenous material that can be used on Gulf television is a major problem.

With a film history that started in the 1920s, Egypt is the paramount media force in the Arab world; it was an early Arab radio, and later, television transmitter and has been a major regional radio broadcaster in the area since the mid-1950s. Records, and then reel-to-reel or cassette tapes, of Egyptian singers, such as Um Kalthoum or Abdul Halim, were prized possessions in the 1950s and 1960s if one lived in the Gulf. With the exception of pre–Civil war Lebanon, the illegal reproduction of records has always been technologically beyond the Arab states. But starting in the 1960s, the widespread use of audiotape in Saudi Arabia helped promote a small pirate music business.[4]

In the 1960s, in major Saudi Arabian cities, one could purchase taped copies of Arabic or Western music from small music sales and duplication shops. Often, one would drop by a shop, examine the selection of albums available and select the one for duplication. The copy would be ready in a matter of hours, often by the end of a shopping trip. This was a fairly low-volume business because the copies were made one by one. Later, high-speed audiocassette dubbing machines made rapid duplication possible. Then, one could go to duplication stores and purchase a cassette tape from a selection already available. By the 1970s, mass copying of music tapes was occurring in Asia, predominantly in Singapore, and in the Arab world. The shops, which at one time duplicated records and tapes, sold them then only because they could be imported so cheaply. "Exports of pirated products from Singapore in 1983 are estimated to include 50 million audiocassettes, shipped primarily to the Middle East, Africa, and South America," according to Wyman (1983). The pirate audiocassette situation in Saudi Arabia alone shows not only the extent of the problem, but the effect it has had, in terms of lost revenue, on the international recording industry. The International Federation of Phonogram and Videogram Producers, a London-based recording industry group that is attempting to stop audio and video piracy, estimates that in 1984, 95 percent of recorded music sold in the kingdom was pirated and that the 60 million units imported that year had a value of U.S.$108 million (*Extent of piracy*, 1985).

The free-enterprise-oriented Arab countries feature even a number of audiotape outlets. In 1985, one of the authors spent several weeks in Kuwait examining media consumption habits. Prerecorded, pirated audiocassettes from Singapore were on sale in music shops for U.S.$1.25 each. The buyers had a choice of famous classical and jazz artists, but popular European and American recordings were most featured. For just over a dollar, one could obtain a selection of the top hits of the 1980s or the latest Egyptian or Lebanese hit records. In 1981, a large Dubai, U.A.E., electronics retailer reported having imported 15 million audiocassettes during the previous twelve months; many were then shipped to other Gulf states ("Mideast entertainment void," 1981). Because of the large expatriate population in the Gulf, audio stores also sold or specialized in recordings of South Korean, Indian, or Filipino artists.[5]

In January 1987, one of the authors surveyed audiocassette stores in Jidda, Saudi Arabia, to determine the approximate percentage of pirated tapes, the cost per cassette, and the range of material available. Virtually all material for sale was pirated, i.e., even Arabic music cassettes were duplicated locally; popular and classical Western tapes were clearly marked as having originated in Hong Kong, Singapore, or Indonesia. The largest audiocassette store in Jidda featured 9,680 separate titles; what with multiple copies, it had a total of 87,120 cassettes for sale. Each tape cost U.S.$3.50; if a customer purchased two, a blank cassette was included. If three were purchased, one recieved a T-shirt advertising the store.

Clearly, when the videocassette boom came to the Arab world, the stage had already been set for the duplication and sale of pirated material.

THE AUDIENCE FOR ARAB CINEMA

> Muslims differ in the degree to which Western civilization attracts them or appears necessary to them: some Muslims attempt to maintain the old customs in a technicalized environment, while others abandon the traditional rules and adopt, to a varying extent, Western ways (Pipes, 1983, p. 107).

British Palestine (during the fifteen years before the end of the League of Nations–United Nations British mandate in May 1948), Egypt, and Lebanon were the centers for feature film showings of both Egyptian products and those imported from the United States and Europe (Samir Seikaly, owner, Philadelphia Cinema, personal correspondence, Amman, Jordan, March 18, 1985). The cinema business in these countries was similar to that in the West at the time—that is, it was composed of large urban theaters that catered to a wide section of the population. In the 1950s, as cities became more populous, more cinemas were built and business was good, since in many areas there was no competition from television.

Naturally, television affected cinema attendance, but only to a minor extent. When the medium went to Jordan, it was initially confined to the capital city of Amman, and the single channel was on the air for only a few hours during the evening. In the mid-1970s, just prior to the introduction of VCRs in Jordan, large numbers of Egyptian workers came there to provide labor for the booming construction business fueled both by investments from the Gulf and by expatriate salaries coming in from Jordanians working in the Gulf. Some downtown cinemas decided to specialize in showing Egyptian films made particularly for a peasant audience, basically the same type of Egyptian working in Jordan. As a result, middle- and upper-class Jordanians concluded that attending the public cinema was no longer acceptable, as it was something Egyptian workers did. By 1982, attending the cinema had become primarily a lower-class leisure-time activity.

In Jordan, the popularity of the VCR only magnified this situation. The desire to acquire a video machine, regardless of one's perceived social class, was heightened by the desire to entertain one's family and friends at home. As will be seen later, VCRs were expensive to purchase in Jordan, but enough Jordanians were working in Gulf countries to bring them back at less expense, even with a hefty customs tax—often 100 percent or more of value—imposed at the airport (Issa Jahmani, director, Press and Publications Office, Ministry of Information, personal correspondence, Amman, Jordan, March 17, 1985).

The film exhibition industry in Egypt has not suffered the same fate as that of Jordan. Sufficient numbers of Egyptian films, particularly those known as "road shows," featuring popular singers, are released each year to guarantee a steady clientele, particularly since cinema ticket prices are government regulated and remain quite inexpensive. Also, the television and VCR penetration rate in Egypt is much lower than in the Gulf states.

The cinema business in the Gulf states developed in a manner quite different from that in Jordan, Egypt, Lebanon, or Palestine. One must remember that the modernization of the Gulf states has been recent, essentially since the 1960s. It was only shortly before Sheikh Zaid bin Sultan, the present ruler of Abu Dhabi, exiled his older brother and took power in 1966, that income from the mounting oil wealth was taken from under the bed of the ruler and deposited in local banks ("Demise of a Midas," 1966). Until the late 1960s, the Gulf countries, with the exception of Saudi Arabia, had formal diplomatic and military agreements with Britain. The British represented these states internationally and provided assistance with security and public administration.

A good deal of the social life of Westerners in the Gulf centered around the established British community of expatriates, who were, quite naturally, interested in what was happening at home. In addition to British books, magazines, and newspapers, some films were imported and shown at embassy functions and various private clubs. However, there were few shown until the late 1940s and early 1950s, a period during which the 16 mm projector became more common, and major film organizations produced prints of releases on 16 mm, rather than 35 mm, or even larger, film stock. This change brought about more widespread use of films in diplomatic and military missions abroad. No longer did one have to be on a large ship or a major military installation to view the latest cinema releases. Soon, projectors made their way into the homes of the wealthier local families.

In isolated areas of the world, films shown by various diplomatic missions have always served as a means of attracting local residents and expatriates to gatherings; the Gulf was no exception to this. Saudi Arabia is a unique example of how 16 mm projectors and films—the antecedents to VCRs—spread and were utilized by the elite.

The origin of the ban on Western gadgetry dates from the reign of King Abdul Aziz ibn-Saud, the founder of Saudi Arabia. Conservative Islam—most particularly the dominant Wahhabi sect of Sunni Islam in Saudi Arabia, which follows the teachings of Mohammed ibn Abd al-Wahhab—interprets the Koran literally. After Abdul Aziz consolidated power in the 1920s, the *ulema* (religious elders) pressed for the status quo to be maintained, while the king urged the introduction of automobiles, telephones, radio (wireless) telephones, and photography because they were needed for the administration of the state. The *ulema* were opposed to Western gadgetry in general, fearing its effect on Bedouin life; but they were most vehemently opposed to wireless communication as the work of the devil because they could not "see" how it worked. The king overcame their objections by having the Koran read into a microphone in Mecca and listened to by the religious men in Riyadh. Abdul Aziz concluded, at the end of the experiment, that the device was the work of humankind and not the devil, because the devil could not pronounce the words of the Koran (Benoist-Mechin, 1958; Eddy, 1963). Eventually, the *ulema* agreed to the use of radio communication, and later, radio broadcasting.

Photography (like television later), however, was a more difficult matter because of the visual element. King Abdul Aziz had to work his magic again in convincing the *ulema* that aerial photography, needed for oil exploration, was compatible with Islam. Eddy (1963) explained:

> Islam takes seriously and literally the Second Commandment to make no graven image nor any likeness of any living thing. Sculpture and painting, impudent attempts to imitate the Creator, are proscribed, sweeping away the idols of the polytheists who flourished in Arabia before Muhammed, but also casting a blight on all the fine

arts so far as live subjects are concerned. And yet, Ibn Sa'ud was convinced by the engineers that photography was essential, especially aerial photography to locate roads and geological foundations. When American photographers began to operate with his permission, he was denounced by bigots for perfidy to Islam.

He summoned his detractors and convened the *ulema* . . . and put forth question: Painting and sculpture are idolatry, but is light good or bad? The judges pondered and replied that light is good; Allah put the sun in the heavens to light man's path. Then asked the King, is a shadow good or bad? There was nothing in the Qur'an about this, but the judges deduced and ruled that shadows are good, because even a holy man casts a shadow. Very well then, said the King, then photography is good because it is nothing but a combination of light and shade, depicting Allah's creatures but leaving them unchanged. The battle was won in the King's characteristic way, by persuasion and not by force (p. 258).

The step from photography to television was comparatively easy. The specifics of how television developed and the effect the medium has had on the kingdom are somewhat beyond the scope of this chapter. However, in order to understand why VCRs are so popular in this conservative Islamic nation, one must know that in the mid-1960s there was strong opposition to the medium and that its introduction led indirectly to the assassination of a popular monarch.[6]

When television was introduced in 1965, it was opposed by reactionary religious elements primarily because of the medium's ability to broadcast material deemed potentially objectionable—i.e., Western programming that contained information inconsistent with Wahhabi beliefs.

Public cinemas remain outlawed for several reasons, most notably because the government feels that it would lose control over what would be shown; a particular concern among religious leaders has been the depicting of women in films. Newspapers, magazines, and books entering Saudi Arabia were censored, and so too would films be. But the possible objections to films, both Egyptian and Western, are so numerous—any references to Christianity and Judaism, alcoholic beverages, inappropriately attired women, kissing, sex, and excessive violence—that little material would be left to show. The censorship problem with films is similar to the dilemma faced by censors in the early years of Saudi television, as described by Boyd (1970/1971):

> The town sheriff walks into a bar—censored because alcohol is forbidden. Sheriff talks to woman who is unveiled—censored because woman's face is shown. Sheriff pets dog as he walks down the street—censored because the dog is considered an unclean animal.

Finally all scenes involving the sheriff are omitted because it is discovered that the sheriff's badge closely resembles the Star of David and is unacceptable because of the association with Israel. The movie was never shown because more film was on the cutting-room floor than on the take-up reel (pp. 76–77).

As previously noted, diplomatic missions showed films in embassies and in the homes of diplomats. The United States Information Service (USIS) library in Jidda allowed Saudis to borrow for short periods for home use, a projector and films, the latter educational in nature and not Hollywood feature offerings.

The Western Province (the Hijaz) of Saudi Arabia, particularly the provincial capital of Jidda, has always had a more Western orientation than the Nejd, the interior of the country. It was in the Hijaz that a well-known Saudi family very quietly started a private film rental business. By the late 1950s, the Jamjoom family of Jidda had a small business whereby families could rent 16 mm projectors and films for a day or two to show at private functions. Of course, the wealthier families, who had projectors at home, rented only films. One could rent American, European, and Egyptian feature films that had been carefully selected for family-type entertainment. The cost was high: about U.S.$50 for both projector and a film for twenty-four hours. The government permitted this activity to take place because the showings were not public; the clientele were upscale, influential people; the renters included high-ranking government officials. The Saudis objected to *public* cinema showings, not to the *private* use of films. The government was against objectionable material such as blue films, but not the family-oriented features that were more common in the 1950s and 1960s than now. Of course, the VCR rental stores killed the film rental business, but the Jamjooms and other film renters moved from film to videocassettes. Visitors to the kingdom, or even those in transit at the three major international airports, may purchase cassettes featuring films from Arab countries, as well as cheaply produced action films from Asia.

The government's liberal attitude toward the Jamjooms' film rental business in the Western Province was motivated in part by the fact that the large Western expatriate work force was already viewing films on a nightly basis. At first, one had to be invited to the British, French, or American embassy in Jidda to see a current film. Then, the American military mission started showing films nightly for the benefit of United States military personnel who, in turn, were allowed to invite a limited number of guests.

Aside from the Dhahran-based Arabian-American Oil Company

(ARAMCO), whose concession allowed film imports for employees, the first American company to have regularly scheduled film showings in the western part of Saudi Arabia was Trans World Airlines (TWA). At the time the airline had an exclusive management contract with Saudia (the Saudi Arabian national airline) and brought hundreds of United States pilots, flight engineers, mechanics, and electronics technicians and their families to the kingdom. Near the airport was built what became known as the TWA Compound, where many families lived, and which was also the location for an American school and the TWA outdoor film theater. It was operated as a club, with eligible members paying dues to see a specified number of films per week. Of course, the airline's main office helped with the film ordering and shipment, something that was made easy, as TWA pilots flying regularly to Europe simply brought back the latest films.

Then, in the mid-1960s, there was a large influx of Americans to Jidda to work for Lockheed and Raytheon, companies that had won contracts with the Saudi Arabian Ministry of Defense. The kingdom offered little for Western workers to do, particularly those without families; and film showings became an important part of entertaining employees at the end of a long workday. The Jidda experience was repeated throughout the kingdom where companies posted workers. Saudi nationals employed by foreign firms could, of course, attend these showings, which soon became the de facto public cinemas in the kingdom. Similar developments were taking place in other Gulf states, but only Saudi Arabia has had a specific prohibition against public cinemas.

TELEVISION SETS THE STAGE FOR VCRS

To say that the introduction of television was a hit in Saudi Arabia is an understatement. At the time, the average Saudi lacked a comprehensive radio service—that which was offered was restricted to a few hours per day—and did not see films. Financially, the mid-1960s were not the super wealthy mid-1970s, but even then, there was a growing middle class that could afford imported monochrome television sets. The financially less well-off acquired sets through the pooling of resources with extended family members. Seven years after television started, Boyd (1972) surveyed citizens of Riyadh about their viewing habits. Some of the findings include:

1. Just over ninety percent of respondents owned a television set; 29 percent owned two sets.

2. Sixty percent of respondents said they watched television daily.
3. Respondents watched an average of 14.8 hours per week.
4. The average number of people viewing with the respondent was 5.9.
5. Ninety-eight percent of respondents believed that television had an impact on Saudi Arabian society, most people noting that it provided information of an overall educational nature to viewers. Respondents also noted that television helped keep the family together.

The same year, the United States Information Agency, through a contractor, conducted its first media survey in Saudi Arabia in order to determine the popularity of international radio stations such as the VOA and the BBC. A few television media-habit questions were included in the survey; findings were similar to what Boyd had found above: 87 percent said they owned a television set and 52 percent said they watched television daily (USIA, 1973). Additional stations and relay transmitters quickly spread television throughout this large country, and by 1980, some fairly remote areas could see the Riyadh-based single national channel by means of low-power transmitters fed by satellite.

THE ARRIVAL OF VCRS

In 1968, a few Saudis personally imported from Europe Sony CV model helical-scan "educational" video tape recorders. These were largely curiosity pieces, used to record family functions. Their fate was sealed by the total lack of maintenance facilities, although, occasionally, the machines were repaired by moonlighting American NBC engineers at the Riyadh and Jidda stations, even though they were hampered by a lack of spare parts.

In 1974, an American working for the Saudi Sony agent (a high-ranking member of the royal family) started selling Sony U-Matic video recorders that use a three-quarter-inch cassette with a maximum playing/recording time of one hour. There were three reasons why the U-Matic models gained only limited popularity. First, they operated on the 525-line NTSC American standard. Sony initially envisioned them for the industrial and educational market in North America and Japan, which uses the American color and line system. Those wanting to use them in the Gulf area required a United States-standard receiver. In fact, the recorders started the custom of purchasing multistandard sets that could receive both the United States and European (625-line system), as well as the three international color systems: NTSC, PAL, and SECAM. Second, there was little prerecorded material available on the one-hour tapes. Even if one wanted to time-shift, one could not do so because of incompatible standards. Third, the U-Matics were expensive. The initial U.S.$6,500 price (Bakhaider, 1981) for machine and monitor did not inhibit the better-off from purchasing them, but it did discourage middle-class acquisition.

What was needed was something more readily available than 16 mm projectors and less expensive than U-Matic machines; that something arrived in the Gulf when Sony introduced the half-inch Beta-format VCR.

The half-inch machines arrived in 1975 and were initially expensive at U.S.$2,500 (Bakhaider, 1981), but not as costly as the U-Matic models, and owners could time-shift because Beta machines used the European standard. Beta machines had additional advantages: they were smaller in size and lighter in weight, featured timers to start and stop recording, and, more importantly, were capable of recording and playback for more than an hour. The cassettes themselves were half the size of those used in U-Matic machines, a factor noted later as important for the smuggling of sensitive films and television programs into the Gulf states. Now an entire film could be placed on a single cassette.

In 1981, a Sony executive reported that of the 2 million VCRs the company had made by then, 20 percent, or 400,000 units, had been sold to dealers in the Gulf ('Mideast entertainment void," 1981). Soon after the introduction of the Beta VCR in the Gulf, JVC, Panasonic, Hitachi, and other Japanese companies began to market VHS-format machines. Several major manufacturers sold VHS machines through local dealers, well organized by the later 1970s to handle the exploding consumer goods market. VHS machine prices were more competitive than those for Beta. By 1980, just as in other developing and developed areas of the world, VHS became the dominant home VCR format. Bakhaider's (1981) survey in the Eastern Province of Saudi Arabia reflects the situation in the Gulf as a whole: 76.6 percent of VCR owners had VHS machines, while only 21 percent and 3 percent, respectively, had Beta and U-Matic models.

THE GULF STATES: A VIDEO CATALYST
FOR THE ARAB WORLD

Excluding North Africa, it is virtually impossible to discuss video in Arab states such as Egypt, the Sudan, Jordan, and Syria without first exploring the flow of both video machines themselves and cassette

TABLE 3.1   PER CAPITA GNP IN THE ARAB WORLD (UNITED STATES DOLLARS)

Country	1978	1982
Gulf States		
Bahrain	$ 4,060	$ 9,860
Kuwait	15,970	19,610
Oman	2,790	6,370
Qatar	15,050	22,060
Saudi Arabia	6,590	15,820
U.A.E.	15,020	24,080
Other States		
Egypt	420	670
Iraq	1,850	15,340
Jordan	1,100	n.a.
Sudan	340	1,690
Syria	960	430
Yemen (Arab Republic)	410	1,680
Yemen (PDR)	450	500
		470

Source: From World Bank. (1980). 1980 World Bank Atlas. Washington, D.C. and World Bank, (1986). World development report 1985. New York: Oxford University Press.

material. Of course, money is the essential ingredient; and when the VCR boom broke, the Gulf states had enormous amounts of it because of oil price increases. They used the revenue to build and purchase everything imaginable, with the aim of fulfilling ambitious development plans. The already wealthy native population became superrich; the middle class gained a measure of wealth; and those in lower income brackets (government clerks, drivers, and low-level administrators) attained the status of middle class, largely through salary increases, the elimination of all taxes, gifts of land for houses, and low- to no-interest loans for home construction. As compared to other Arab countries, actual wealth, expressed in GNP per capita (World Bank, 1980, 1985), gives some indication of both the need for expatriate workers and the motivation for them to work in the Gulf (see Table 3.1).

There are other ways of looking at these data. The Gulf states have relatively small populations with insufficient manpower to staff the armed forces, government, schools, clinics and hospitals, and the technically oriented jobs in the oil-based economy. The poorer Arab countries—e.g., Egypt and Jordan—have little oil and most of that is consumed domestically.

Before the riches arrived in the 1970s, there already existed the custom of hiring expatriates for many jobs. Skilled workers, such as computer operators, teachers, and physicians, came primarily from Lebanon, Jordan (a large number of Palestinians with Jordanian passports), and Egypt. Construction laborers, as well as some drivers, came from Egypt and North Yemen. The Lebanese were often involved in private business as managers and administrators.

By 1983, 80 percent of the labor force and 60 percent of the population in Kuwait consisted of expatriates. At the same time, 90 percent of the U.A.E. work force was foreign (Choucri, 1986). Almost everyone came for the money rather than for the challenge or the culture. There seemed to be funds for every project imaginable. After all, the sale of oil by the Gulf states to the petroleum-starved industrialized countries realized an income that could not be spent as quickly as it was generated. By 1982, the dollar flow to the Gulf amounted to the largest international transfer of wealth in history.[7]

When an Egyptian or a Yemeni works in the Gulf, he[8] goes with the intention of staying for a few years and amassing a nest egg in order to purchase an automobile or an apartment, or in general, to provide a better life for his family. In 1975—the beginning of the large increase in the Gulf work force—it was estimated that 28.1 percent of the Jordanian work force was employed abroad. Similarly, 3.8 percent of the Syrian, 3.2 percent of the Egyptian, and 20.3 percent of the North Yemeni work force were employed outside their respective countries (Birks & Sinclair, 1979). Most of these workers went to the Gulf states, and their number increased significantly after 1975.

Because of the technological dermands of the construction in the Gulf, fewer laborers from the less-developed Arab countries were brought there after 1980. No longer were buildings made only from standard concrete and steel; newer building methods required a skilled work force that was unavailable in the area. Also, South Korean contractors were starting to outbid even local firms for contracts. The South Koreans provided their own workers, who were flown in on chartered aircraft. Yet even with the large influx of Asian workers in the Gulf, tens of thousands of workers from other Arab countries were needed. Egyptians, Yemenis, and Jordanians had both the requisite skills and a command of the language.

Aside from the indigenous market for videocassette recorders, merchants shifted their marketing strategy to cater to Arab expatriates, as well as to those from Asian countries. Both the actual sale of machines and the sale and rental of cassettes were involved.

During periodic trips home, workers are accustomed to taking back gifts. Most of these have been small home appliances readily available in the Gulf at reasonable prices, but both scarce and expensive in the Sudan, Yemen, or Egypt. Portable radios with cassette players were popular in the early 1970s. In the mid-1970s, small portable black-and-white television sets were carried on flights from the Gulf to these countries. From the later 1970s until now, VCRs have been the popular item.

Egyptians working abroad often live together in cramped quarters with the aim of minimizing living costs. As a way of occupying leisure time, they share appliances. Almost everyone can afford a personal radio and cassette player, but individual ownership of a television receiver and VCR is economically difficult if one's aim is to take home as much money as possible. So, one person will own the television set and another the VCR, on which are shown rented tapes of Egyptian films. The VCR owner usually collects a designated monthly amount from his roommates for the use of the machine.

When visiting home, an Egyptian worker will either take a VCR back to his family, sell it at a profit to friends, or sell it to an Egyptian retail store. In Egypt, most appliance dealers are not exclusive agents for one particular electronics manufacturer, but rather handle a variety of brand names. A good sample of Gulf-purchased VCRs for sale in Egypt can be found on 26th of July Street in Zamalek, a district in Cairo. Stacked on the sidewalk are boxes of VCRs purchased in the Gulf, brought through airport customs, and immediately sold to retailers. The Egyptian price is usually about three times the cost in the Gulf. At between U.S.$1,000 and $1,200 per machine, only a small percentage of the Egyptian upper class can afford a locally purchased imported Japanese or South Korean videocassette recorder. In 1984, Egypt started assembling VCRs from kits imported from Japan. Although somewhat cheaper, the Egyptian-made machines have a reputation for being less reliable than the imported ones (Jehan Rachty and Mohamed Wafai, Faculty of Mass Communication, Cairo University, personal correspondence, Cairo, Egypt, January 4, 1987).

As World Bank figures indicate, Egypt is an economically disadvantaged country; yet after more than twenty-five years, television broadcasting is widespread wherever electricity is available. Data from 30,000 interviews commissioned by the Egyptian Ministry of Information in urban and rural areas produced the following results:

1. There are 4,252,942 television sets in Egypt; 13.6 percent are color sets.
2. The average number of viewers per set is 6.4.
3. Sixty percent of the Egyptian population views television.
4. Television viewing averages 3 hours and 49 minutes per day. The most popular viewing times are between 7 and 8 P.M. and between 10 and 11 P.M.

5. Television is viewed most frequently on Sundays and Fridays, the respective Christian and Muslim days of rest (Middle East Advisory Group, 1983).

In rural areas, it is a status symbol to own a television set, particularly a color set. Items such as television sets show that a male family member, the patriarch or son, has "made it" economically and is providing the very best for his family. After color television receivers appeared in rural homes, VCRs followed. A videocassette recorder was an even more impressive symbol of economic success than a television set. Soon, small video rental shops appeared in rural areas to supply material for the VCRs. This situation was repeated in other rural areas in Syria, Jordan, Iraq, and Yemen.

VCR OWNERSHIP IN THE ARAB WORLD

Even if VCR sales figures for the area were available, they would not be helpful in determining machine penetration rates because, as previously noted, too many are purchased with the intention of export. The percentage of homes with VCRs is more easily learned from the survey data that are available. In 1982, Kuwait's Pan Arab Research Center, a Gallup affiliate, undertook a study on behalf of Japanese VCR manufacturers to determine brand and format preferences, as well as to learn something about how people used VCRs. The research was done in both Kuwait and Saudi Arabia and indicated that more than half of all homes with television sets also owned VCRs.

> In 1982 52.7 percent of Kuwaiti homes with television sets owned VCRs; just over 20 percent owned more than one machine. 1981 was the most important year for VCR purchases—28.2 percent said they purchased units during this year (Pan Arab Research Center, 1982, October).
> In 1982 61.4 percent of Saudi Arabian television homes owned VCRs. About 17 percent owned more than one machine. Also, in Saudi Arabia 1981 was an important year for VCR purchases—27.6 percent said they first acquired machines this year (Pan Arab Research Center, 1982, November; see also Pan Arab Research Center, 1986).

In late 1984, the Pan Arab Research Center conducted another survey; results show that the percentage of ownership increased dramatically. In Kuwait, 85 percent of television homes had VCRs; and it was anticipated that by 1985, the market would be saturated (Sami

TABLE 3.2   1986 ESTIMATED GULF STATE VIDEOCASSETTE RECORDER OWNERSHIP

Country	% of TV Homes with VCRs
Kuwait	88%
Saudi Arabia	75
Bahrain	79
Qatar	77
U.A.E.	80
Oman	75

Raffoul, manager, Pan Arab Research Center personal communication, Kuwait, February 9, 1985).

Bakhaider (1981) found in his survey of videocassette recorder usage in the Jidda, Saudi Arabia, area that all 120 respondents owned machines. In 1986, al-Oofy (1986) studied VCR ownership patterns and usage among high school and college students in three Saudi Arabian cities: Riyadh, Medina, and Jidda. Seventy-six percent of male and 83 percent of female respondents in the survey said they or their families owned a video recorder; those without one at home said they had access to one at a friend's house. In Gulf homes, a VCR is nearly as common as a refrigerator.

Based on available survey data, discussions with media experts and personal visits to Gulf states, estimates of the percentage of VCR ownership among citizen of six Gulf states were made, as reflected in Table 3.2.

Estimating the VCR penetration in Arab countries outside the Gulf is more difficult, primarily because so many machines enter illegally or because so many are brought in legally, but not counted by customs. Also, we are unaware of any academic studies or marketing surveys done in the non-Gulf Arab world. However, discussions with Ministry of Information officials and personal observation have produced a reasonable assessment of VCR ownership. (See Table 3.3.) Three countries—Yemen Arab Republic, Lebanon, and Jordan—are examined specifically.

Yemen Arab Republic

North Yemen is a beautiful, isolated country situated south of Saudi Arabia on the southwestern tip of the Arabian peninsula. Because of the limited economic prospects in Yemen, many Yemenis work in the

TABLE 3.3   1986 VIDEOCASSETTE RECORDER OWNERSHIP

Country	*% TV Home with VCRs*
Egypt	4%
Iraq	10
Jordan	30
Lebanon	55
Syria	10
North Yemen	25
Sudan	1

Gulf. Most workers take on semi-skilled and unskilled jobs as domestic servants, drivers, office cleaners, and laborers.

There are three ways of acquiring a VCR in Yemen. First, one can purchase a legally imported machine, an expensive undertaking reserved for only the very wealthy. Second, one can bring a VCR back from the Gulf, paying customs duties at the airport. Third, one can purchase a VCR smuggled into Yemen from Saudi Arabia. Since the end of the Yemen Civil War in the late 1960s, there is a history of goods illegally entering the country from Saudi Arabia. First, large trucks filled with goods from the Red Sea port of Jidda make their way south to the Yemen border. From then on, they are loaded, transported, and unloaded by a tribe that over the years has become expert at this undertaking. Goods are then placed on smaller, four-wheel drive vehicles for transport through the rugged terrain of northern Yemen to major cities such as San'a and Taiz. At least one-third of the VCRs in Yemen have entered in this manner. In 1984, the BBC surveyed three urban areas in Yemen to determine international and domestic radio listening preferences. The survey instrument also included questions about VCR ownership. The data show that 26 percent of television homes in the three largest urban areas had a VCR (G. Mytton, head, BBC External Service, International Broadcasting and Audience Research, personal communication, London, January 23, 1985).

VCRs are, of course, concentrated in urban areas where there is electric power, but anecdotal evidence suggests that the machines have spread in some rural areas to homes with generators. Further, there is evidence that in addition to the standard cassette offerings of Egyptian and Western films, a great deal of Asian and European pornography is available and is being seen by women when men are not at home.

Lebanon

It is difficult to imagine a war-torn country such as Lebanon still functioning after what it has endured since the civil war began in 1975. Not only does Lebanon function, but some sections thrive despite the street fighting and what appear to be constant car bombings. There have been occasional disruptions in the radio and television system; residents often experience electricity failures, and those who can afford them have purchased their own generators. The video market flourishes in Lebanon because people seem to want some relief from the constant anxiety of daily life.

Initially, VCRs were expensive, but after most government functions broke down after the civil war started, competition and smuggling lowered prices (Abboud, 1986). Video rental stores operate there in a similar manner as those in the West or the Gulf. Reportedly, there is great demand for children's programming, as parents want to keep children out of trouble and out of their way, while providing them with something educational.

Jordan

After losing the West Bank to Israel during the 1967 war, Jordan suffered economic hardship and political instability, culminating in the 1970 civil war between various Palestinian factions and forces loyal to King Hussein. When the Gulf oil boom started in the mid-1970s, Jordan was in a position to benefit. The country was attractive for some Gulf investment, and the huge Jordanian expatriate work force in the Gulf sent back large sums of money. This lifted a stagnant economy and fostered a construction boom. Because Jordan is not a wealthy state, it imposes taxes upon citizens, including a particularly high import tax on clothing, appliances, automobiles, and electronic equipment.

Videocassette machines may be purchased locally, but they are expensive. Usually people working in the Gulf bring them back during yearly home leave, saving at least U.S.$100 over the cost of local purchase. In 1985, the VCR penetration for Jordan was estimated to be 25 percent (Issa Jahmani, director, Press and Publications Office, Ministry of Information, personal correspondence, Amman, Jordan, March 17, 1985); by 1986 that figure was presumed to have reached at least 30 percent.

VCR REGULATION

The existence of large numbers of VCRs, particularly in the Gulf states, is both a concern and an embarrassment to some governments. Despite efforts to control software distribution, most governments have decided that there is no way of stopping this almost universal home appliance. There are those who either deny the pervasiveness of VCRs in some areas or, in response to concerned governments, seem to ignore their popularity. For example, in 1985. Three Arab communication researchers and writers published a UNESCO-sponsored monograph dealing with the history, needs, and priorities of mass communication in the Arab world. Nowhere in the sixty-page document is reference made to the home videocassette recorder (Bakr, Labib, & Kandil, 1985).

Previously mentioned was the initial lack of concern about VCRs on the part of Arab governments. Either those few who saw the potential negative effect of home recorders did not voice any opposition, or their opinions fell on deaf ears. Many Arabs have taken quite literally the statement by Saudi Arabia's founder, Abdul Aziz ibn-Saud: "our faith and your iron" (Eddy, 1963), meaning, "we will retain our Islamic faith, but we must have Western technology." Until the late 1960s, several Gulf states did not rely on Greenwich Mean Time (GMT). Saudi Arabia, for example, used "sun time" officially. This is based on the Islamic, specifically Wahhabi, idea that it is midnight when the sun sets each evening. Of course, the sun sets at a different time each day and it sets at different times in different parts of Saudi Arabia. There existed elaborate charts for converting sun time to GMT and vice versa. For a while, dual-faced watches were popular because they could refer to both GMT and sun time.

Saudi Arabia eventually converted to GMT in the late 1960s for two main reasons. First, the government realized that if it was to join the modern world, it had to fall in step with other countries. Broadcasting, airline, and international telecommunication traffic had to be synchronized with other states, including her neighbors. Second, even before the 1974 oil boom, a fine watch was a prized possession. In the major cities of Jidda, Riyadh, and Dammam-al-Khobar, the Rolex, Omega, and Piaget agents did well. Those who could not afford such expensive timepieces quickly acquired the lower-priced high-tech watches introduced in the late 1960s by Citizen and Seiko of Japan. Third, the ordinary Gulf Arab eventually asked for conversion to international time standards. Technology does have an impact on culture, particularly if its messages prevail against existing political and religious values.

The home videocassette recorder is not a minor household convenience; it is a device of major social significance that allows owners to view what they want, when they want. If broadcast television is mass transit; then the VCR is the private automobile. Home video also provides a means of circumventing censorship imposed by virtually all Arab states on television programming and films.

As the concern about the subject matter being seen on VCRs began to arise in 1980, various Arab states established procedures for reviewing incoming material. However, few specific standards existed for the censorship of imported films. Saudi Arabia had a particular problem, because no formal mechanism existed for reviewing imported films at ports of entry, since public cinemas do not exist. Films on Saudi television were censored before being telecast. In the Gulf, Bahrain recognized the problem and initially assigned its film censor committee to deal with it. In December 1979, when the committee began to review imported cassettes, its chairman realized that the task would be impossible for the four high-ranking government officials. Because the group viewed about 300 films per month, the inclusion of tape would have added an unreasonable burden ("Censors to check video tapes," 1979). Eventually, the Ministry of Information was given the responsibility for cassette censorship. Video shops are now required to provide a list of offerings, and MOI officials regularly monitor what is being offered by cassette outlets (Isa bin Rashid al-Khalifa, Undersecretary, Ministry of Information, personal communication, Manama, Bahrain, August 26, 1984).

Generally, the more conservative the country and the higher the percentage of VCR penetration, the greater the desire of the government to control available cassette material. The is particularly true of the Gulf states.

A great deal of videocassette monitoring is informal and is carried out by the ministries of information. The exception to this in the Gulf is Oman, where responsibility for controlling video material rests with the Ministry of National Heritage and Culture. The problem is that it is impossible to stop effectively the importation of material thought by governments to be undesirable. To illustrate the point, the measures taken by Saudi Arabia to regulate video shop are provided.

By the late 1970s, the kingdom's rulers, realizing the potential problems of unrestricted video imports, created legislation, the Video Bill (see Appendix 3.A), to deal with undesirable material. Rules governing video came into force in February 1980; they were to be enforced by the Ministry of Information. The legislation provides specific guidelines for those eligible to own and operate a video store: citizenship and age (the owner must be a Saudi citizen over age 18 and not a student); location (the store must be on a main road with a visible entrance, and not near a mosque); and license (including a security check by the intelligence service). The regulations further stipulate that women are not allowed to enter video rental or sales stores. The bill omits specifics of what one may not rent, but it does prohibit "material contradicting [the] Islamic faith or morals or adversely affecting the security of the country" (Bakhaider, 1981, p. 114).

What is prohibited on Saudi television or within the private film rental business has always been rather general. One gains some insight from Ministry of Information guidelines for imported material from the late 1960s:

1. Scenes which arouse sexual excitement.
2. Women who appear indecently dressed, appear in dance scenes, or in scenes which show overt acts of love.
3. Women who appear in athletic games or sports.
4. Alcoholic drinks or anything connected with drinking.
5. Derogatory references to any of the "Heavenly Religions."
6. Treatment of other countries with praise, satire, or contempt.
7. References to Zionism.
8. Material meant to expose monarchy.
9. All immoral scenes.
10. References to betting or gambling.
11. Excessive violence (Shobaili, 1971, pp. 242–243).

In 1982, Saudi Arabia enacted a new comprehensive information policy aimed not only at television and videocassettes, but meant to address many internal and external information concerns; the policy was expressed by a body known as the Higher Council for Information. A group of high-ranking government officials headed by the Minister of the Interior, the council is empowered to deal with a running conflict within the government over information policy. One particular part of the policy deals with television and is divided into four sections: religious, social, political, and technical. Selected areas of concern outlined by al-Aamoudi (1984, Appendix B) include:

1. Religious: Ban, a. everything that opposes or offends the Islamic religion whether directly or indirectly; b. all types of preaching and advocating of other religions; c. the customs and rituals of people other than Muslims; d. betting and loans with interest; e. the show[ing] of alcoholic drinks or people drinking or referring to it in the dialogue; f. scenes or words about pork and bacon, statues, religious pictures, and all nude statues.
2. Social: Ban, a. all that conflicts with the social and behavioral

values; b. bars, clubs, and all sorts of fun fairs; c. scenes of drug taking; d. dancing except folklore and national dancing in decent clothes; e. kisses, embracing, exciting love scenes, and adultery.
3. Political: Ban, a. all that contradicts or offends the government or rulers; b. all political and party principles and slogans that contradict the country's policy; c. strikes and demonstrations, acts of vandalism, spying, political and military plotting; d. all that points to racial discrimination.
4. Technical: Ban, a. all films and videos that are technically not fit to be shown because of distortion of the picture; b. shows of ridiculous stories or those which are boring; c. weak artistic production whether it is in acting, producing, or writing.

The above policy may indeed be the correct one for Saudi Arabia, as it does address the concerns of the *ulema*. However, the strict adherence to these, and to guidelines imposed prior to 1982, makes for dull television, as the Saudi audience response in surveys show. Such policies only increase the desire to see banned material, much of which is available on cassettes, either directly from rental shops or from the video underground.

As previously mentioned, there were several reasons for the kingdom's information policy. Although privately acknowledged as impossible to enforce, the video censorship provisions were designed to keep various types of undersirable tapes out of circulation. Of particular concern among officials was *Death of a Princess*, a film made for television in 1980. Supposedly the reenactment of the late-1970s execution of a female member of the Saudi royal family for the crime of adultery, the film was shown in April 1980 on commercial television in Britain and then, a month later, on public stations in the United States. The Saudi government strongly objected to the film on the grounds that it was not an accurate depiction of Saudi culture or the crime itself. At one point, the Saudi government asked the United States Department of State to stop the scheduled telecast. The Saudi government's objections drew the attention of the international press to a film that otherwise might have gained little notice. Naturally, Saudis, as well as other Arabs, were interested in the film. The Saudi Arabian Video Bill addresses government concerns about video rental shops.

The night *Death of a Princess* was first shown in the United Kingdom, it was taped directly off the air and flown the next morning to Saudi Arabia's Eastern Province, where it was duplicated and distributed. Twenty-four hours after the film was televised in the United Kingdom, Saudis were watching it in their homes (Boyd, 1982). It was available in Gulf capitals within days of being telecast in Britain ("Mideast entertainment void," 1981).

Not all Gulf states have specific rules for video stores. For example, it is relatively easy to regulate the seventy-five video retail stores in Bahrain because it is a small country. The Ministry of Information requires that video stores provide a list of tapes, and ministry officials regularly examine what is being offered by stores (Isa bin Rashid al-Khalifa, Undersecretary, Ministry of Information, personal communication, Manama, Bahrain, August 26, 1984). Kuwait has a similar system. But Saudi Arabia presents a different problem because of its size, location, and wealth. Arab countries want to keep out undesirable material they believe to be either against the prevailing political system, anti-Islamic, or pornographic. However, as the *Death of a Princess* situation shows, this is impossible.

The number of video outlets in Egypt is still quite small. The authorities there are more concerned about pirated tapes of Egyptian films than about the availability of undesirable Western material. Egypt, however, has had the equivalent of Saudi Arabia's *Death of a Princess* affair. The American made-for-television film *Sadat* was banned by the Egyptian government because the Ministry of Culture felt that it did not reflect the former president's role in Egyptian politics after 1952. They also objected because Sadat was played by a black actor, Louis Gossett. Not only did the Egyptian government ban the film from public showing in Egypt, it also prohibited Columbia Pictures, the distributor, from showing any of its films in Egypt (Miller, 1984). When the film was shown on American television, it was taped and flown to Egypt and other Arab countries, where it freely circulated. Of course, the ban by the Egyptian government only drew attention to a film that otherwise might not have been publicized; *Sadat* has been seen by a substantial percentage of the educated Egyptian upper class—those with access to VCRs and the most likely to have an interest in the film.

Egypt is a country in which the distribution of cassette programming is especially difficult to monitor, particularly in Cairo and Alexandria, where the underground economy thrives. For example, it is possible for a small family to survive in Cairo on income the husband receives from collecting trash from residences, selling facial tissues or cigarettes to motorists in stalled traffic, or shining shoes. the Egyptian video boom has produced yet another underground occupation: renting videocassettes door-to-door. The Arabic word for satchel—a cloth, wicker, or plastic shopping bag—is *shanta*. The *shanta* man is someone who regularly visits appartment houses to offer a variety of cassettes; for a fee of approximately U.S.$1, he leaves a tape for twenty-four hours (Yousef Shaheen & Hussein Fahmy, personal communication, Cairo, Egypt, January 5, 1987). Some *shanta* men specialize in pornography, old or new Egyptian films, and, for areas of the city where

Westerners live, American, British, and French films and television programs.

Like every other Arab country, Jordan was slow to recognize the effects videotape would have on its media and society. The problem of control is addressed by the Ministry of Information and other agencies. In addition, cassettes entering the country are identified at the Amman airport or at Saudi Arabian, Syrian, or Iraqi border-crossing points by customs officials, who either allow them to enter or hold them until they can be viewed by members of the Censors' Council, a body functioning under the Ministry of Information's Press and Publications Office.

In order to operate a video rental establishment in Jordan, one must have a license from the Press and Publications Office. As of March 1985, there were 430 shops in the country, four having been closed during the first quarter of 1985 for renting "undesirable material." Despite its best efforts, the Censors' Council cannot view all of the videocassette material legally entering Jordan. Between 1982 and 1985, the four members of the council reviewed 20,000 cassettes. Criteria for rejecting cassettes are sexual activity unacceptable to an Islamic society, excessive violence, antireligious material, and films and television programs that depict Israel in a positive light (Issa Jahmani, Director, Ministry of Information, Press and Publications Office, personal communication, Amman, Jordan, March 17, 1985). Officially, the Ministry of Information is concerned about pornographic and other types of undesirable material, but officials realize that they are essentially powerless to stop it from entering the country (Michael Hamarneh, Undersecretary, Ministry of Information, personal communication, Amman, Jordan, March 17, 1985). Jordanian officials will continue to try to stop the importation and distribution of what they believe to be unacceptable video material, but privately, they admit that almost everything imaginable is available on tape and that virtually nothing the government can do will alter the situation.

Oman has attempted to stop non-Omanis from entering the country with videotapes in their luggage. On the application for an Omani visa are stamped restrictions on what one may bring into the country; in addition to liquor, one may not import any videotapes, including blanks.

VIDEOCASSETTE PROGRAMMING

As stated previously, virtually everything is available in the Arab world on videotape, some of it of little interest to Arabs. In addition to North Americans and Europeans, there are substantial numbers of Pakistanis,

Indians, Thais, South Koreans, and Filipinos in the Arab world, mostly in the Gulf states; and they have brought with them taped programming from their own countries. Also, some local videotape rental stores cater to Asian workers by importing material.

The poorer Arab countries are not the only states to have benefitted from the Gulf oil boom. It was estimated that in 1975 there were 154,000 Indian workers in the Arab world; in 1982 the number was 913,000. In 1979 the number of Filipinos and South Koreans was estimated at 80,000; by 1981 the number of Filipinos had climbed to 342,300 and the number of South Koreans to 182,400 (Choucri, 1986). When many of these Asian workers returned home, they took with them videocassette recorders. There were three phases to the pirating of videocassette material in the Arab states, and they are discussed in the following sections.

The Informal Importation of Western Material

In the early VCR years, people returning from Europe and Asia would bring films and television programs with them. For the most part, these were for the owner's personal use, but whatever entered a country was shared with family and close friends. About 1979, when the number of machines increased and those with the financial means started acquiring second VCRs, copies of taped material were made by linking two VCRs. Two friends or family members would get together for an evening of tape viewing and duplication. With second or third copies of master tapes in circulation, the availability of imported material greatly increased. During this phase, the availability of material was the paramount concern; the technically poor quality of second- and third-generation copies was tolerated, but people soon demanded a better-quality picture.

The Organized Pirating of Western Material

In the Gulf, local businessmen with European and North American connections started to acquire video material because they were able to make agreements with local contractors (with large numbers of expatriate employees), hospitals, and the numerous first-class hotels springing up for the business traveler, to supply them with videotaped programming. At first, the firms arranged to have television programs in Britain, films from pay cable services such as Home Box Office (HBO), and television shows in the United States taped directly off the air and shipped to the Arab world. In 1980, a BBC spokesman said, "We learned of one organization which taped the entire BBC output every evening and flew it off by private jet next morning to an Arab country" ("Thieves who steal your TV shows," 1980). Although United

States television shows such as "Dallas" are preferred, the absence of commercials on the BBC and the limitations on advertising breaks on Britain's commercial channels made the United Kingdom a favorite pirate market.

At first, little attention was given to editing the programs; it is still not unusual in the Gulf to watch a videotape of a television program and see a portion of the evening news of an earlier date from London or New Orleans. This happens when a pirate tapes programs later in the evening and lets the machine run through the late evening news. One sees less of this in the mid-1980s because video companies pay more attention to editing. Also, Western television programs are used less frequently now; organizations needing video material usually contract with local rental establishments for recent films.

Particularly before rental stores became common in the Gulf, pirated American television programs were used in some hotels. Many of the larger hotels that belonged to well-known international chains used material legally obtained. However, many first-class establishments under local ownership and management continue to use pirated material.

The following schedule from the Ramada Inn in Doha, Qatar (presented here only to show the pervasiveness of United States television programs and not to imply that they were illegally obtained) details those United States programs shown on the internal video channel on Saturday, January 9, 1982 (*Ramada News*, 1982):

    3 P.M.: "Children's Hour"
    4 P.M.: "Name of the Game"
    6 P.M.: "The Autobiography of Miss Jane Pittman"
    8 P.M.: "Rockford Files"
    9 P.M.: "Dying Room Only"
  11 P.M.: "Alfred Hitchcock"

This hotel and others in the Gulf states feature a rotating schedule of television shows. In addition to those noted above, the Ramada Inn also showed the following American television programs for the week of Saturday, January 9, 1982: "Quincy," "Emergency," "The Bionic Woman," "McMillan and Wife," and "Baretta." Of course, whatever was available to hotels was also for private rental.

In Riyadh, Saudi Arabia, the King Faisal Specialist Hospital operates its own internal television channel for the convenience of patients as well as the large resident foreign staff. For the week of March 10–12, 1979, the United States programs used were: "Quincy," "McMillan and

Wife," "Switch," "Harper Valley PTA," "M*A*S*H*," "Starsky &
Hutch," "WKRP in Cincinnati," "Alfred Hitchcock," and "Ironside"
(*KFSH TV Guide*, 1979). The hospital had its own guidelines for editing,
which conformed to those then in effect at the Ministry of Information.
However, most of the programs shown on the internal television channel did not appear on national television. A program such as "WKRP
in Cincinnati," for example, would be challenging to edit for a Saudi
audience.

Thus far the discussion has centered on the pirating of Western
films and television programs. Although this piracy continues, Western programming has a limited appeal in this Arabic-speaking region,
except to Western-educated or Western-oriented persons. Beginning in
the early 1980s, Arab world video shops have featured Asian Kung-fu-
type films. Many of these cheaply made features come to the Gulf from
Hong Kong via Singapore, an Asian video pirate capital. They feature a
great deal of violence, and the action-adventure orientation appeals not
only to the large number of Asians but also to Arabs. One does not
need to know English—the language into which most of the films are
dubbed—in order to follow the story line. Such offerings are often
seen on the internal video channels of second-class hotels in Jordan
and the Gulf.

The Organized Pirating of
Arabic-language Video Material

Most Arabic-language television programming and films available on
cassette are Egyptian. Egypt has long dominated the market for both
film and television in the Middle East. This does not mean that other
countries, such as Lebanon, Syria, and Iraq, do not produce desirable
material. Unfortunately, Lebanon's internal strife since 1975 has all but
eliminated sizable numbers of films and television programs available
for sale elsewhere. Syria and Iraq have only small film industries, as
they lack the talent and money to undertake more than a handful of
productions per year. Since 1980, however, both countries have increased considerably their taped television output.

Egyptian feature films are paramount in the region, and there is
great competition for the latest releases on videotape. Even Egyptians,
to whom first-run features are widely available in local theaters, clamor
for pirated tapes of the latest films. Although this is nearly impossible
to stop, the Egyptian government is attempting to do something about
the illegally obtained and distributed films. The government has
understandable financial concerns: state-organized film companies are
losing money and income is lost through decreased cinema ticket sales.

One of the main differences between the conditions applying to the VCR industry in the Gulf states and in Jordan is the long-established Jordanian cinema business. In the 1950s several family businesses with theaters in Lebanon and British Palestine moved to Jordan and built theaters. Some specialized in Egyptian films; others showed mostly American and British films with Arabic subtitles. When television came to Jordan in the late 1960s, theater attendance decreased somewhat, but families at all economic levels continued to go to the cinema because it offered material not then available on the nation's single monochrome television channel. By 1980, however, theater attendance was down dramatically, primarily because of the proliferation of VCRs (Samir Seikaly, owner of the Philadelphia Cinema, personal correspondence, Amman, Jordan, March 18, 1985).

Starting in 1983, some theater owners in Amman decided that they must do something to compete with the shops supplying rental cassette material. Most theaters had fallen into a state of disrepair because income had decreased. Some owners went into the video rental business, refurbished older theaters, and built new ones. The modernized Rainbow Theater in the old section of Amman, and the newly constructed Philadelphia Cinema in one of Amman's newer areas, are examples of what was done to increase cinema attendance. Some of the new or modernized theaters are multipurpose facilities and can be used for live bands, plays, and concerts; yet competition from VCRs may prevent them from surviving. Pirated material is on the market before first-run Egyptian and Western films are in theaters, and owners are essentially powerless to do anything about it. Although the following example deals specifically with Jordan, it illustrates a problem found throughout the Middle East, where theater owners are in competition with VCRs.

The Seikaly family has been in the cinema business since 1943, moving from Palestine to Jordan in 1948. The family were only distributors of film until 1954, when they built theaters and became exhibiters. Business was good until 1978, when the VCRs first arrived; and in 1981, the Seikalys themselves also got into the video rental business, albeit temporarily.

To get started in the video rental business, they purchased and imported 10,000 blank VHS and Beta tapes at U.S.$15 each. Six months later, the price of tapes dropped dramatically to $6. Because they lost so much money on the blank tapes and because video rentals were so competitive, the family sold its shops, deciding to return to operating theaters. The new U.S.$1.4 million Philadelphia Cinema is the Seikalys' most recent and most expensive investment in Jordan. The family believed that they could make money by exhibiting films that could not be obtained on cassette.

During the first part of 1985, Samir Seikaly signed a contract with an Egyptian film company for the exclusive Jordanian rights to an Egyptian film entitled *Huna Qahira* (*This Is Cairo*), starring Mohammed Sobhi, a popular actor and singer. The contract stipulated that if unauthorized versions of the film were to appear in the Jordanian market prior to March 15, 1985, the distributor would pay the Seikaly Cinema Company U.S.$50,000. This action was taken by Seikaly, because he suspected that some Egyptian film companies sold the VCR rights at the same time they sold the theater rights, thus making double the normal fee.

On March 2, 1985, the film cleared Jordanian airport customs and was released by the Ministry of Information censorship office to Seikaly. Despite locking the film in a vault, Seikaly found that videocassettes of *Huna Qahira* were immediately available in Amman's video rental stores. How the film reached the stores is a matter of speculation. Seikaly questioned customs officials to determine whether they had allowed someone to copy it, and he additionally questioned ministry officials. The mystery remains unsolved, although Seikaly believes that someone in the film-processing laboratory in Cairo made a cassette of the film and sold it. The Philadelphia also booked a Western film entitled *Body Rock* for an exclusive showing, but again pirated videocassettes were on the market, this time two weeks before the scheduled showing (Samir Seikaly, owner, Philadelphia Cinema, personal communication, Amman, Jordan, March 18, 1985).

The Jordanian video revolution is also financially affecting the government. The average price of a single theater ticket in Amman for a first-run film in one of the new cinemas is U.S.$4.80. Of this amount, about half is government tax. Videocassettes are inexpensive to rent, particularly compared with the cost of a theater admission. Several plans are available through the hundreds of video stores in Jordan. The standard scheme involves joining a club for about U.S.$20, which entitles the member to several film rentals. Thereafter, a cassette costs approximately U.S.$1.50 for 24 hours. Following the practice in many Western countries, competition is driving rental prices down. Some rental shops no longer require an initial membership fee. If the tape is of something worth keeping, a copy may be made by using the VCR of a friend or relative.

There is a curious side to the pirating of Western films on videocassette in the Arab world. If a popular offering in its original version either offends the viewer or is objectionable to a government, the meaning of the film or television program can be changed through subtitling. Traditional subtitling of feature films is expensive, because after the appropriate dialogue is determined, it is optically printed on film prints to be exhibited. Starting in the mid-1970s, Arab television stations started electronic subtitling. This was occassionally done with cooperation of companies selling American or British television programs, because an already subtitled program was more marketable. Jordan is still the leader in the Arab world in showing popular United States shows such as "Dallas." Jordan Television received a price break for showing this program first; an additional incentive for showing it was that it was rented at an especially low rate with the understanding that Jordan Television would subtitle the program and allow the dialogue scrolls to be rented to other Arab world television stations. Most Western programs are telecast with live subtitles—i.e., with the dialogue "keyed" over the videotape. This is done by using an electric typewriter to place the desired Arabic dialogue on a white scroll. When the program is aired, the subtitling is done by a machine that uses a monochrome camera with the polarity reversed, so that the black letters on the white scrolls appear white when keyed over the picture. A bilingual operator sits in a booth during the telecast, listens to the soundtrack, and advances the subtitles on the scroll according to the prepared script. Because this is done live, it can be somewhat sloppy, particularly if the operator gets too far behind or ahead of the relevant scene. However, the mistiming is not obvious to those viewers who speak only Arabic.

The pirate cassette of the American film *Rambo—First Blood* circulating in Syria has taken on a rather different meaning after video distributors subtitled it in both French and Arabic to conform to Syrian government wishes. The film tells the story of an American returning to Vietnam to rescue United States prisoners held by the Viet Cong. Syria is closely associated with the Soviet Union (its main arms supplier); and possibly in an effort not to offend the Soviets, the Syrian government decided to change the action from Vietnam to World War II. When one of the actors says in the film "You made a hell of a reputation for yourself in Nam," this is changed in the subtitles to, "You made quite a name for yourself in Guadalcanal" (the scene of a famous World War II battle between United States and Japanese military forces) (Robinson, 1985, p. 6).

The newest twist to Arab video, particularly in popular first-run Western and Egyptian films, is advertising. In an area where television advertising is either limited, not allowed (as in Oman), or only just starting (as on the Saudi Arabian second channel), those wanting to display cigarettes, watches, automobiles, cosmetics, packaged food products, and home appliances, have found a new outlet. Not all new releases have advertising, but it is not unusual to see commercials for Marlboro cigarettes and Citizen watches crudely edited into tapes rented from video shops.

Beginning in late 1985, an unusual alliance—Egyptian film producers and directors, wealthy Saudi Arabian video shop owners, and Gulf-based advertising agencies—started producing films in Egypt for two express purposes: (1) to attract new videocassette renters with fresh material from well-known artists, and (2) to expose this audience to heavy doses of consumer advertising. The producer of such material thus receives revenue from the usual daily rental fee, as well as from the commercials inserted into the film.

It works like this: A representative of a Saudi Arabian video distribution company, such as Jamjoom, contracts with an Egyptian film producer and director for exclusive rights to a film. The production usually stars a well-known, but still aspiring, young artist. After the film is "rough-cut," it is dispatched to Saudi Arabia, where the final edit is done and commercials—as many as twenty or thirty—are inserted. For financial reasons, virtually all films are put on two-hour, rather than three-hour, cassettes. As of 1987, the Egyptian government and concerned film producers were attempting to stop this practice for several reasons, the most notable of which concerns a 1986 incident. It seems that an Egyptian singer had agreed to star in a Saudi-financed "made for video" film with the proviso that a specified number of his songs be included in the film. This was agreed to, but when the final video-cassette of the film was released for rental, not one of his songs was included. It seems that because of the two-hour cassette, the songs had to be removed in order to make room for commercials (Yousef Shaheen and Hussein Fahmy, personal communication, Cairo, Egypt, January 5, 1987).

REVOLUTIONARY AND POLITICAL VIDEO

It has already been noted that banned films circulate freely, although unofficially, on videotape. Governments attempt to censor material they believe to be morally offensive or hostile to them. Increasingly, however, material considered to be political, rather than cultural, in nature is being seen in Arab homes with VCRs. Some tapes of religious and political discussions circulate on videotape among a small group of people interested in learning about opposition political movements.

These programs are usually crudely done and present little more than people talking directly to a home video camera.

In the Gulf States, on the other hand, cassettes of professionally produced satirical plays are in demand because they extend the reception of this Arab art form to a wider population. Abdul Hussein Abdul Ridr is a Kuwaiti producer, writer, actor, and comedian best known in the Gulf as a political satirist. His biting humor is popular in Kuwait, where the Ministry of Information subsidizes some of his productions. Perhaps his best-known production is *Bye Bye London*, a satirical comedy about Gulf Arabs visiting Britain. Allowed to circulate freely in Kuwait, it is banned in Saudi Arabia, where it is available on the underground video tape market.

Similarly, *The Keeper of the Keys*, a musical featuring the Lebanese actress and singer Fairouz is available only on videocassette because it contains an unmistakable political theme:

> ...when the greediness and oppression of "the King" becomes excessive, the people decide to leave the country and they go—all but one woman, Zaid al-Khair, to whom they entrust the keys of their houses. With all the people gone, the King realises the futility of his greed and love of power and begs Zaid al-Khair to ask them to return to their country ("Middle Eastern story-tellers," 1985, p. 53).

VIDEO PORNOGRAPHY

It is always difficult to write about something that is taboo in a society, because it is difficult to document how it is used. Pornographic films and video material are the subject of strict regulations in many Western countries. Blue films on videotape are forbidden in virtually the entire Arab world. Some countries, such as Jordan and Egypt, are more liberal than the Gulf states, but obscene movies are not in keeping with the teachings of Islam, no matter how liberal the ambience. However, some pornography on videocassette is present, even in the conservative Gulf. It is readily available for purchase in European and Asian international airports, and some people bring it through airport customs. Undesirable video material also comes to the Gulf the same way contraband has arrived for hundreds of years—by sea. Private video companies also distribute pornographic material through video stores. Some shops provide this type of material openly; others are more circumspect. In the Gulf, those wishing to rent a pornographic tape usually ask for one at a shop and receive a tape marked with the name of a well known nonpornographic film. The store tells the renter or purchaser that the tape has a false start—that is, the tape does begin with the film marked on the label, but at a designated number on the VCR tape counter, the actual pornographic film starts. This system helps avoid detection should a tape be examined at customs or in occasional spot checks of video stores by Ministry of Information officials.

> In the United Arab Emirates (UAE) some shops selling what the government sees as offensive material are closed regularly for a period of three days or more, while films with more than 10 sex scenes—usually lingering kisses and fondling—are confiscated. Earlier this year, the UAE government also stopped giving licences to new shops, saying the 90 or so video shops operating in the Emirates had saturated the market. In Kuwait, diplomats say most of the taped material seized by customs officials is pornographic ("Gulf crackdown on porn," 1986, p. 110).

Some residents of Kuwait have complained to the Ministry of Information about the amount of pornographic material available on the market, but privately, officials we interviewed are unperturbed. The occasional rental of pornographic material and subsequent embarrassment is illustrated by an incident involving an Arab family living in Kuwait. In early 1985, the family rented a cassette of E.T. to show at their four-year-old child's birthday party. By mistake, the shop gave the family the pornographic E.T. The tape started suitably for the children with the opening scene from the actual popular American movie, but then changed several minutes later to the pornographic version. The video shop supplying the tape told the family that the two versions had been switched accidentally in the shop. In a 1985 study conducted in Kuwait by the Arab States Broadcasting Union's Arab Center for Audience Research, it was found that 18.5 percent of respondents said that they watched video material for "sex thrills." This does not mean that the material was necessarily pornographic, but it indicates that films and television programs with romantic and sexual themes are popular and have a potentially "great negative effect on the general values in the society" (Adwan, 1985, p. 15).

The blue film has been found in the Arab world since 16 mm film projectors made the viewing of such material possible. However, VCRs and consequent access to pornographic material on cassette are causing worry among most governments and some individuals. No firm evidence exists at this point, but it is now believed that pornography is being viewed by women in North Yemen, as previously noted, and by adolescents in the Arab world.

VCR TECHNOLOGY

The story of how home video technology developed in the region is really the story of how Japanese manufacturers quickly understood, and adapted to, the Arab VCR market. The first VCRs sold in the Gulf were European-standard machines, and the majority recorded and played PAL color, the system used in Gulf countries, except Saudi Arabia. When color transmissions began in the Gulf, television and, later, videocassette owners were faced with the need to purchase multistandard television receivers and VCRs if they wanted to see all of the television channels available, as well as to view tape in the various standards circulating with pirated material from Egypt, Europe, and North America. To reiterate, in 1987, someone living in the Eastern Province of Saudi Arabia could see at least thirteen different channels. When color was first started in the area in the mid-1970s, our hypothetical Saudi viewer needed a three-system television receiver in order to see color signals from the one government channel (625-line, SECAM); from the Arabian American Oil Company (ARAMCO) (525-line, NTSC) station[9]; and from neighboring states—all 625-line, PAL. Of course, many people purchased single-system sets, but as color and VCR development in the area occurred after 1974, many people could afford the luxury of the multisystem, multicolor receivers. There was, and still is, high status attached to having one.

With a sizable number of multiline multicolor sets in the area, it naturally followed that the major Japanese manufacturers would market VCRs to complement these receivers. Those desiring to view the vast offerings of mostly pirated material from the United States, Britain, and Egypt needed the flexibility of the multisystem VCRs. Typical of machines for sale in the region was JVC's HR-7600 MS, a VCR capable of recording both PAL and SECAM signals off-air and playing tapes from five systems: PAL, SECAM (Europe), MESECAM (the Middle East), and two NTSC standards, 3.58 and 4.43 Mhz (Victor Company of Japan, n.d.). In Oman, this and similar machines cost approximately U.S.$500 to $600, but PAL-only units with a wired remote control are available there for about U.S.$350. In Kuwait and Saudi Arabia, the same machines cost about 15 percent less because of increased retail competition and the fact that there are no import duties.

Japan's keen reading of the Arab market has resulted in an impressive sales record, particularly in the Gulf states. In 1985, one manufacturer introduced a new VCR (a double-cassette video recorder) that seemed ideal for the Arab world. The machine, capable of duplicating tapes by itself, and reported to have been manufactured for an electronics retailer in Saudi Arabia, was not well received in Western countries, where film industries and governments alike are attempting to stop video piracy. The double-cassette VCR is an ideal device for the Arabian Gulf states, and its development points to the sensitivity of the Sharp Electronics Company and its Middle Eastern distributors to customer wishes in the Gulf Market (Watts, 1985).

THE IMPACT OF VCRS

At this early stage of development, it is difficult to make a final assessment of the impact VCRs have had on the Arab world. The medium is too new; too little research has been done on the phenomenon, and reaction is still taking place. However, some preliminary conclusions can be drawn by examining video's effect on broadcast television, on the cinema production and exhibition business, and on the audience.

Television

The effect of video on television varies dramatically from one section of the Arab world to another. We assume that the impact is greater as penetration increases. In Egypt and the Sudan, the low rate of VCR ownership means that viewers are not being pulled away from the national channels to cassette offerings. In fact, recent evidence from a Norwegian anthropologist (Wikan, 1985) working among Cairo's poor, indicates that television set ownership among this group has grown substantially since the late 1960s, and that the medium is having a positive effect on some of the poorest families in this crowded urban area:

> The three TV sets listed [for the 1969 study] were recent acquisitions (1968, 1968, and 1969 respectively) and had only recently changed the daily pattern of life in those households. Now everyone shares in the enjoyment TV introduces into a life of urban poverty, where cinema visits are at best rare and memorable life experiences and holidays, vacations and even birthday celebrations are empty words only. The entertainment introduced into such lives by Egyptian TV comedy and soap opera genuinely changes the level of living. . . .
> With the advent of TV in the home, family coherence suddenly increases. Husbands who previously spent their free time in cafés, now prefer to stay at home instead, and that in turn entails greater

family commitment on his part; it leads to a greater appreciation of the home, and therefore to a much greater readiness to contribute financially to it (pp. 12–13).

A large-sample survey, done in1983 by the advertising department of Egyptian television, indicated that about 60 percent of the Egyptian population watches television (Middle East Advisory Group, 1983).

Surveys done by Jordan Television also show that television is popular, but the survey—like the one noted above for Egypt—was done for advertising sales purposes and does not ask questions about VCR viewing nor preferences for the neighboring Syrian television channel. However, those who advertise on Jordan's highly commercialized channels are apparently concerned that they will no longer reach the audience they once did. Those in charge of Jordan television fear that it will become more difficult to provide programming that will keep people in front of their television receivers, viewing one of the national channels rather than watching programming on videocassette. Chapter 1 described the programming decision of a Jordanian television official who cancelled an ABC film when he realized his audience had already seen it on cassette.

In the Gulf states, the VCR penetration figures noted earlier speak for themselves. At least in these states, according to studies, one of the major attractions of home video is as an alternative to government-run television. Adwan's (1985) VCR study found that video usage took time previously spent on other media; 22.5 percent, 23.3 percent, and 22.4 percent of respondents in Iraq, Kuwait, and Qatar, respectively, said they watched less television after having purchased a home video recorder.

Cinema Production and Exhibition

A few films are produced in Iraq and Syria, but Egypt is the Hollywood of the Arab world, and video has had a negative effect on this important Egyptian export industry. The story of the Egyptian film *Huna Qahira* in Jordan was noted earlier. It is symptomatic of what continues to occur in other Arab countries. In the Gulf, for example, people can enjoy a first-run Egyptian film at home at a fraction of the cost of seeing one at a cinema; often, no cost is involved if a pirate tape is borrowed from a friend or family member. Cinemas have become places where workers from Asian and other Arab countries watch films, and cinema attendance now implies that one does not have a VCR. Respondents in the Adwan (1985) study said that they attended film showings less often after having purchased a VCR. Specifically,

TABLE 3.4   VCR UTILIZATION PER WEEK IN THREE ARAB COUNTRIES

Viewing Hours Per Week	Iraq	Kuwait	Qatar
1– 4	12%	37%	51.5%
4– 8	26.5%	34%	7.5%
8–12	27%	9%	12%
12–16	25.5%	11%	10%
No Response	9%	9%	9%

Source: From Research on video programs in Iraq, Kuwait, and Qatar by N. Adwan, 1985, Baghdad, Iraq: Arab Center for Audience Research.

31.6 percent in Iraq, 31.7 percent in Kuwait, and 14 percent in Qatar noted that VCR ownership had this effect.

Western film distributors have essentially given up on the Arab world, because they know that either copies of their releases will find their way to local video rental stores when a film opens, or a pirated film will be in shops before the legally acquired film is exhibited, having entered the country through Singapore. According to Roy Featherstone, president of CIC International, which markets MCA and Paramount films outside the United States, "By releasing our video titles in [the Arabian peninsula and East Asia], we are in essence granting licences to steal" (Schiffman, 1984).

Audiences

It is no longer possible for Arab world governments to ignore the fact that material seen on home video machines is having some impact, possibly negative, on viewers. However, before a discussion of video's influence on the audience is possible, it is desirable to review some findings on why, and under what conditions, people view taped material.

Number of viewing hours. The data presented in Table 3.4 from the Adwan study (1985, p. 17) show video viewing hours per week for three countries. Further, the survey results indicated that the number of VCR viewing hours per week was higher for younger people and for females.

Bakhaider (1981, p. 42) found in his study of the Jidda, Saudi Arabia, area that a high proportion of VCR owners *watched video a great many hours each week* (see Table 3.5).

TABLE 3.5   VIEWING HOURS PER WEEK IN JIDDA AND BAHRA

Hours Per Week	Total Number of Responses
20–26	22
14–19	37
6–13	30
2– 5	10
Not Available	21

Source: From The impact of the video cassette recorder on Saudi Arabian television and society by B. Bakhaider, 1981, unpublished master's thesis, San Diego State University, San Diego, CA.

In al-Oofy's (1986) study of VCRs in three large Saudi Arabian cities, the majority of both male and female secondary school and college students said they watched video material two hours per day. This is similar to the number of respondents' broadcast television viewing hours.

Popularity of VCR programming. Two reasons for video usage stand out. First, cassette programming can be turned on at a time to suit the viewer and can be interrupted, while broadcast television ties the viewer to the set and is continuous. Second, VCR viewing helps fill leisure time. Of course, leisure time can be filled in a number of ways, but Adwan (1985) states that in the three countries he studied, video tends to dominate leisure hours and has decreased the amount of time spent on studying, reading, and consuming the broadcast media. Dissatisfaction with what is available on television is a paramount reason for video programming use. This does not mean that television is neither desirable nor important, but television may be becoming primarily a purveyor of news and of local and international sports. In areas where there is large VCR ownership, broadcast television as an entertainment medium seems to be decreasing in importance. This has interesting implications for all of the Arab world, but most particularly for the conservative Gulf states. There, television systems were created and programmed to provide viewers with material that is consistent with traditional Islamic values and that does not promote Western ideas, offer excessive violence, or show sex scenes. Sex is a particularly taboo subject, yet Adwan (1985) found that among VCR owners surveyed in Kuwait, 18.5 percent indicated that video material was used for "sex thrills" [nonpornographic scenes found in films and Western television shows] (p. 15). Bakhaider (1981) found that 65 percent of his respondents said they were attracted to video programming because Saudi television's only channel at the time lacked new and interesting programming. Thirty-five percent of the respondents said they used video to fill surplus leisure time.

Both male and female college and university students in al-Oofy's (1986) study gave as a first reason that they or their families owned VCRs, the lack of alternative forms of entertainment. For females, the second most important reason for VCR ownership was the limited amount of television programming. For males, the second most important reason was a lack of quality television programming.

VCR programming preferences. As with previous topics in this section, there are few studies on which to base observations. Yet, in 1985 and 1986, one of the authors examined video stores in Egypt, the Sudan, Jordan, Saudi Arabia, Kuwait, Bahrain, and Oman and found Egyptian films; Egyptian television shows, mainly soap operas; Western films; Western (mostly American) television programs; Asian Kungfu films dubbed into English; and Indian, Filipino, Thai, and South Korean material imported for expatriate workers.

One's preference for cassette programming of a particular national origin largely depends on education—including a facility in English— and Western orientation—i.e., travel to and familiarity with Europe and North America. Fifty-one percent of the respondents in Bakhaider's (1981) survey preferred Egyptian programming—films and cassettes of music and dance—and only 28 percent rented Western films and television programs. Some of the Egyptian programming available in rental shops has been shown on Saudi television. However, on television such offerings are censored; on cassette, one may see the uncut version. The most popular type of material rented was the thriller (either a mystery, suspense, or horror movie), followed by love stories and variety programs such as comedy acts and programs featuring singing and dancing.

In 1986, al-Oofy found a preference among his male and female Saudi respondents for Western, specifically American, shows; Egyptian material was ranked second. The most likely explanation for the different preferences in the two Saudi studies is the educational level and age of respondents. Al-Oofy's sample consisted of secondary school and college students. The most popular VCR material among the males in al-Oofy's study were thrillers. The females preferred comedies and musical varieties.

Effects on Viewers

Even in societies where social scientists have done a great deal of research, the findings on the effects of media are largely inconclusive

**TABLE 3.6   THE IMPACT OF VCRS ON THE UTILIZATION
OF OTHER MEDIA**

Activity	Iraq	Kuwait	Qatar
Less radio listening	10.0%	13.5%	17.3%
Less television viewing	22.5%	23.3%	22.4%
Less cinema attendance	31.6%	31.7%	14.0%
Less theater attendance	20.1%	14.6%	9.1%
Less reading	13.8%	12.7%	19.1%
No response	2.0%	4.2%	18.1%

Source: From *Research on video programs in Iraq, Kuwait, and Qatar* by N. Adwan, 1985, Baghdad, Iraq: Arab Center for Audience Research.

and, at best, debatable. The only known survey in the Arab world of the effects of VCR viewing was al-Oofy's survey in three major Saudi Arabian cities. Respondents were asked whether VCR viewing had affected their style of speaking, their manners, the way they dressed, and their beliefs and values. Some males and females admitted to being influenced by video material, but the majority said that they were largely unaffected or that the effect was "average." But when asked whether they thought that VCRs represent a *danger* to traditional Saudi Arabian society, the majority said that there was a strong, or very strong, danger.

> The respondents are very uncomfortable with the negative influence that VCR programs would have on Saudi attitudes, manners, beliefs, and values. Both genders in the questionnaire in all cities express anxious signs toward the videotapes.
>   Figures show that 34 percent of male respondents believe in a strong negative impact of the VCR on Saudi culture and 52 percent of the females believe in an even stronger negative impact.
>   Among these negative symptoms, subjects refer to violence, consumerism, and immorality (al-Oofy, 1986, p. 90).

The effects that VCRs have had on the consumption of other media are more easily determined; at least some information is available from Adwan (1985). The Arab States Broadcasting Union's Baghdad-based research center asked respondents in Iraq, Kuwait, and Qatar whether VCR viewing caused them to alter consumption of other media. The results are shown in Table 3.6.

Regardless of the specific effects on Arab viewers attributed to videocassette recorders, the machines ultimately will be judged to have had some impact. In those areas where VCR penetration is high, viewers have changed their consumption of other media, particularly state-run electronic media. At least with regard to films and television programs, it is the Arab viewer, rather than the program producer and broadcaster, who is now in charge of what is seen.

NOTES

1. Originally these countries were known as Persian Gulf states because they are on what geographers have termed the Persian Gulf. During the years they were under British protection, they were known as the Trucial States. However, the Arabs prefer "Arabian" to "Persian" for obvious reasons. This became so particularly in the late 1960s during the British withdrawal when the ever touchy relationship with Iran became more strained over disputed oil rights in the Gulf.

2. Videocassette recorder (VCR) refers to video machines that both record and play material. Originally VCRs were just that, devices allowing owners to see pre-recorded material as well as to record programs from television or cable systems. Recently manufacturers began marketing videocassette players (VCPs) which only play material. In some developing countries, the VCPs may prove successful because they are less expensive. However, in the Arab states—particularly the Gulf—the VCR is the standard tape machine because there is so much home recording of borrowed and rented material. Also, in the high-income Gulf, the cost of machines is not a major factor determining whether residents purchase them.

3. When Lebanon was founded in the early 1940s under terms agreed to by France and the League of Nations, it was agreed that the population was equally divided between Christians and Muslims. However, because of a higher birth rate, Muslims have come to be the majority. This was only one of the problems leading to the 1975–1976 Civil War.

4. Although the word piracy originally referred to "The practice or crime of robbery and depredation on the sea or navigable rivers, etc. . .," *The Oxford English Dictionary* notes that the term was applied to the theft of intellectual works, "Literary Piracy," as early as 1808 (*Oxford English dictionary*, 1961, pp. 900, 901).

5. In 1985, one of the authors attempted to purchase a Julio Iglesias tape in a Kuwaiti music store only to be told that the Ministry of Information would not allow them to be sold because the artist has been banned by the Damascus-based Arab League Boycott of Israel office. (The singer apparently had been judged to be a supporter of Israel.) However, the store manager showed the author a large selection of Iglesias tapes that were kept under the counter, because the ministry made occasional checks for banned material.

6. Just prior to the Riyadh television station's air date in 1965, Prince Musad, a royal family member, convened a group of conservative religious supporters

and attempted to march on the station and destroy the equipment. The action was discovered and stopped by the police. Musad was shortly thereafter killed by a policeman under mysterious circumstances. Musad's family appealed to King Faisal to punish the policeman, but Faisal ruled that the action was warranted.

In late March 1975, ten years after the incident, Prince Musad's brother, Prince Faisal ibn Musad, shot and killed King Faisal during a public reception in Riyadh. *Newsweek* reported that the assassin shouted, "Now my brother is avenged!" after the shooting ("The murder of King Faisal," 1975, pp. 21–23).

7. Internationally, oil is priced in United States dollars.

8. This personal pronoun is used intentionally here and elsewhere in this chapter. With the exception of female teachers and medical personnel, the expatriate work force in the Gulf is entirely male.

9. In November 1976, ARAMCO modified its monochrome transmitter to broadcast NTSC color; then in March 1979, the station converted to 625-line PAL color.

APPENDIX

3.A: The Video Bill February 1980

The Ministry of Information
The Kingdom of Saudi Arabia

Saudi Arabia is one of the best examples of a modern videotape society. A wealthy, conservative Islamic state, the kingdom now operates two national television channels. One of the strong attractions that television held for the government when it decided to introduce television in the mid-1960s was that this form of broadcasting offered an alternative to the invasion of foreign, mostly Western, material via film. The country still does not permit public cinemas to operate. By 1980, when video tape recorder ownership had reached approximately half the native population, the Ministry of Information became increasingly concerned about the control of incoming videotape material. The 1979 Iranian Revolution had made the Arabian Gulf States, particularly Saudi Arabia, sensitive to what people were watching and who was watching it. The following details how the kingdom hoped to curb the distribution of cassettes.

Article 1
This bill is intended to regulate the purchase, recording, sale, and rental of video tapes and cassettes by stores in Saudi Arabia. Anyone who has opened, or who wishes to open, a video store is now required to apply for a license from the General Directorate of Publications.

Article 2
The following conditions must be fulfilled to qualify for a license. The applicant must:

    a)   be a Saudi Arabian citizen
    b)   be at least 18 years of age
    c)   have no criminal record
    d)   present at least an elementary school certificate
    e)   *not* be currently enrolled as a student
and f)   intend to be employed full time in this profession.

Article 3
Any establishment dealing in video tapes and/or cassettes must:

      a) be located on a main road and have an entrance that is clearly visible from the road

      b) *not* be an apartment in a residential building

      c) *not* be near a mosque

      d) have a clearly visible, uncovered display window

      e) *not* allow women to enter

  and f) display a valid license and official certificate of the store's ownership,

Article 4

To obtain a license, the owner must supply documentation (including photographs) giving proof that the conditions in Articles 2 and 3 are being met. Besides this documentation, the application for a license should include the following:

    a) Six photographs of the applicant

    b) two photographs of his identity card

    c) a photograph of the applicant's/company's contract or ownership certificate including the names of all others in charge of the store

    d) a document certifying the applicant's blameless conduct (lack of a police record)

    e) a chart showing the store's situation

    f) the applicant's permanent residential address, post office box, and telephone number.

Article 5

After the application has been submitted to the General Directorate of Publications for review and verification, it will then be forwarded to an intelligence agency for a security check.

Article 6

The business must be opened within six months of its licensing.

The period covered by each license is five years, after which time the above conditions will again be checked for renewal.

The Ministry of Information has the right to cancel a license at any time without mentioning its reasons for doing so.

Article 7

The video store license prohibits the owner from practicing any other activities at his establishment except those provided for by the license.

Article 8
The owner of a licensed video store is required to keep a list of his employees, their nationalities, their job titles, and photographs of their residential and working licenses. He should provide the Publication Department with a photo copy of this list and notify it of any changes that may occur. He should have this list ready to be presented to concerned officials when they ask for it during their inspection tours.

Article 9
The owner, or his representative in case of the owner's temporary absence, will be held responsible for any activities or misconduct performed on the store's premises.

Article 10
The owner's license is *not* transferrable and any violation of this rule will result in the license's revocation.

Article 11
The ultimate goal of the circulation of any films and tapes should be the benefit of the viewer and the dissemination of culture, knowledge, and innocent entertainment.

Article 12
The General Directorate of Publications and its branches will censor both foreign films and tapes sent to it by customs authorities, and also all video materials that have been recorded locally at licensed places before they have been presented for use. The recorded materials will be approved for use after it has been ascertained that they contain nothing that would transgress the moral code and/or principles of Islamic law and nothing that might threaten the security of the country. The Directorate confiscates and destroys any films that have not been approved, or any unacceptable part of a film providing the rest is considered suitable for viewing.

Article 13
The owner of a licensed place must keep a list of the films used in his place and approved by the General Directorate of Publications or its branches.

Each film will be stamped with the store's stamp, a model for which is provided by the Directorate.

Article 14
The use and recording of films that contain material contradicting Islamic faith or morals or adversely affecting the security of the country are prohibited.

Article 15
The unlicensed sale, rental, and recording of films and tapes is prohibited.

Article 16
Hotels, hospitals, companies, clubs, restaurants and all other public places screening films and/or tapes are prohibited from showing materials not approved by the Publication Department.

Article 17
Licensed establishments are prohibited from having: (a) loud speakers which cause people discomfort; (b) colored or dim lights; (c) inner rooms; (d) exciting pictures (inside or outside); and (e) curtains which cover the display window. They are also prohibited from allowing women to enter, and from being open during prayer time as well as beyond the prescribed time during the night.

Article 18
Advertisements in local newspapers—as well as any other kind of publicity—about video films, tapes, and establishments are prohibited from being published.

Article 19
Anyone who practices any of the activities mentioned in Article 1 without a license will be subject to a fine of up to 5,000 riyals and his store will be closed immediately. A repeat violation is subject to a fine of up to 30,000 riyals.

If anyone is found with films or tapes violating the censorship regulations defined by this bill, the store will be closed immediately and he will be punished according to Articles 21 and 22.

Article 20
A licensed owner who rents or sells unapproved films or tapes found *not* to contain materials violating censorship regulations will be fined up to 5,000 riyals and his store will be closed for a period not exceeding

two months. A repeat violation constitutes not only a fine of up to 10,000 riyals, but also the termination of the business establishment and the owner's license as well as the confiscation of all unapproved films.

Article 21
Anyone found with unapproved video films that contradict the principles of Islam will be fined up to 10,000 riyals.

An owner who both sells and shows such films will be subject to a fine of up to 50,000 riyals and his establishment will be closed for no more than two months. In the case of a repeat violation, the fine will be doubled, the store closed, the license terminated, and all materials confiscated.

Article 22
Anyone showing any film or tape that includes material either defamatory to the Islamic religion or endangering the security of the country will be fined not less than 100,000 riyals. His establishment will be closed, his license terminated, and the Ministry of Information will refer the matter to the Ministry of the Interior for further prosecution and a possible prison sentence.

Article 23
Anyone who violates Article 16 will be punished according to the seriousness of the violation.

Article 24
Anyone who violates Article 17 will be fined up to 2,000 riyals and/or his store will be closed for not more than two weeks. In the case of a repeat violation, the fine will be as much as 10,000 riyals and/or the store will be closed for up to two months. In addition to the fine, the license could also be terminated if the type of the violation that is repeated warrants a more severe punishment.

Article 25
Violations not mentioned above will be refered to the Minister of Information who will determine the type and degree of punishment.

Article 26
Designated officials of the General Directorate of Publication have the authority to inspect all such establishments, to seize its materials in the

case of an uncovered violation, and to report their findings to the Directorate.

Article 27

Violations will be reviewed by a committee headed by the Director General of Publications and including the membership of a legal advisor and a senior official of the Publication Department chosen by the Under-Secretary of the Ministry of Information. This committee will define all punishments and have the authority to call for testimony from whoever it deems such testimony will be useful before giving its judgement.

With the approval of the Under-Secretary, the committee's decision is final if the punishment consists of not more than two thousand riyals and/or closing of the establishment for not more than two weeks. If the punishment decided by the committee exceeds these limits, the papers will be presented together with the committee's decision to the Minister of Information for his consideration. The decision is to be conveyed to the violator who then has the right to appeal to the Minister within ten days. The Minister's decision will be final.

Article 28

Those currently practicing any business activities mentioned in this bill have three months to apply to the Publication Department to be considered for a license. Anyone who fails to comply will be punished accordingly and his establishment will be closed.

Article 29

Within a month of the issuance of this bill, all video store owners are required, first of all, to present the Publication Department with a list of the film titles they have in stock and, secondly, to turn over all films which transgress these censorship regulations. If any such objectionable films are found on his premises after this one month period has expired, the owner will be punished by law and his application may be refused or his license revoked.

The Publication Department will eventually censor all video films currently displayed on the market, and this work will be performed in the shortest possible time.

Article 30

The Publication Department will present the governor of each region with a list of the licensed video stores within his jurisdiction.

Article 31
The bill of regulations is to be conveyed to whom it may concern and takes effect from the date of its issuance.

The Minister of Information

CHAPTER 4

Videocassette Recorders in Asia

Diversity is the characteristic that best identifies the Third World portion of Asia. A continent that represents about two-thirds of the world population, Asia includes mammoths such as China and India, with nearly one-half of the earth's people, as well as relative dwarfs such as Hong Kong and Singapore, which, as cities, qualify as large, but as states, somewhat small. Asia encompasses all of the world's major religions (in fact, it gave birth to some of them); many of its languages, concentrated in numerous multilingual nations (the foremost of which is India with hundreds); and most of the possible political philosophies.

Economically, Asia includes some of the largest and most successful capitalist, socialist, and communist systems. It includes countries that, given their economic realities, do not belong in the Third World. The countries of Asia range from very poor states, such as Bangladesh, to more affluent ones, such as Hong Kong, Malaysia, and Singapore. Geographically, the continent contains a sample of everything: the highest mountains on earth and nations composed of thousands of islands, such as Indonesia and the Philippines.

The governments of Asia include United States-style democracies, parliamentary democracies, military dictatorships, communist states, and monarchies, as well as mixed systems. Instability has marked politics in many countries with military or emergency rule, dynasties, and one-person or one-family power concentrations. Colonial rulers in Asia included those of Asian origin, such as Japan in recent times, as well as North American and European ones. Most major nations of the West have had their stay in Asia. Colonialism's end is recent, occurring mainly after World War II.

All these factors have had their effect upon Asian broadcasting. In fact, such diversity has spurred the development of radio and television as unifying forces. Colonialists started some of the first broadcast services; their marks are still felt. Former British Asia still follows the BBC model; and the Philippines, colonized by the United States in the twentieth century, patterns its system after that country. National broadcasting services usually are controlled by government departments and, in some cases, public corporations. The Philippines and, to a lesser degree, some other countries have had private ownership of broadcasting. Privatization of broadcasting, a worldwide trend, has moved into Asia, especially Malaysia.

Because they are national in scope, broadcasting services strive to serve developmental purposes. This is natural since development journalism had its birth in Asia in the 1960s. It is often difficult, however, to distinguish between developmental broadcasting and governmental propaganda. The importance of broadcasting can be gauged by the tight controls most governments exercise over it.

Dependency is a characteristic of much Asian broadcasting. Television programming has strong injections of shows from the United States, Japan, Britain and Australia, among others. On the other hand, some Asian countries such as Japan, Taiwan, and Hong Kong have been significant exporters of television programs to their neighbors. Technologically, some broadcasting systems are very dependent on outside corporations, most notably Brunei Television and Indonesia's Palapa satellite.

Perhaps some systems are breaking out of this dependency relationship, as governments and broadcasters set up regional agencies for mutual benefit. Training and other professionalization activities are being encouraged by groups such as the Asian Institute of Broadcasting Development in Kuala Lumpur, CEPTA-TV in Singapore, or the Asian-Pacific Broadcasting Union. These and other regional organizations, such as those set up in ASEAN, have also worked on regional news and program exchanges that are beneficial to broadcasting services.

For years, radio was considered to be the most important and pervasive medium in the region. More and more, television is taking on that role, despite its relative youth. While some countries' television services, notably those in the Philippines, date to the early 1950s, others, such as those in Nepal in 1984, and in Sri Lanka and Brunei in the 1970s, are very new (see Lent, 1978, 1982).

THE ARRIVAL OF VCRS

Home video, for most of Asia, is predominantly a 1980s phenomenon. It arrived on the heels of import liberalization policies, as in India and Sri Lanka; of modernization plans (especially in telecommunications), as in China and Malaysia; or of a waning interest in often dull developmental television programming. In Nepal, home video arrived before national television; in Sri Lanka, simultaneously with television's advent. In virtually every instance, home video's blossoming just seemed to happen; it was not planned.

Among the early adopters of home video were the Taiwanese. In 1970, one of the Republic of China's three television networks, the China Television Company, showed a program on Japanese wrestling that became an instant success. Soon, imported videotaped shows of wrestling, variety shows, and pornography were the main fare of many teahouses, cafeterias, and motels, which showed them to promote business. When the television services banned Japanese wrestling in 1971, it made little difference to devotees since hundreds of eating places already offered it on video. Also, the showing of explicit sex and violence, banned on television, attracted patrons to the video cafés.

Video moved into Taiwanese homes after 1976, when Sony introduced Betamax and National brought out VHS. Video shops that dealt in VCRs and cassettes were established throughout the island, despite a five-year ban in 1982 on all imports (including VCRs) from Japan, the main supplier of video products. The ban, coupled with a rising demand for VCRs, prodded local manufacturers, Taitung and Sampo, to enter the market in 1982. But Japanese-produced VCRs continued to be preferred and were smuggled into the country; Taiwan-manufactured brands were one-and-a-half times more expensive because of the limited market for them and the heavy taxes placed on parts imported from Japan. The total number of VCRs grew 57.5 percent between 1981 and 1984 (Wang, 1986, p. 336).

The liberalization of the import policy by the government that came to power in 1977, and the development of a national television service two years later, led to the initiation of home video in Sri Lanka. Home video's growth was linked to the development of video clubs between 1981 and 1983. The clubs imported cassettes that they rented to members who paid deposits of between U.S.$25 and $38.75 per cassette. By 1983, twenty clubs were functioning in Colombo and others were sprouting in outlying towns; their patronage was between 2,500 and 4,000 members. This form of entertainment was limited to the wealthy few because a VCR cost about U.S.$600 to $1,000 (Kurukulasuriya, 1983, p. 68). In Nepal, home video preceded, and probably pushed authorities into developing, national television. Elites in Kathmandu had been purchasing VCRs while abroad to record and play Indian television programs and foreign movies, which prompted officials to develop a television service in December 1984 (interview, Subhadra Belbase, program director, Worldview, Singapore, November 21, 1986).

The development of television in India had been a model of self-sufficiency for other Third World countries—until the Asian Games of 1982 in New Delhi. At that time, the government decided to promote the medium, with the aim of giving the games a larger audience. Initially, the authorities decided to import 90,000 color television kits for local assembly and sales. But soon after, the customs duty on preassembled sets was drastically reduced; over 150,000 television receivers flooded into India as the receiver price dropped from U.S.$5,000 in 1981 to $800 two years later. The VCR was similarly popularized with the lowering of duties. Before the 1982 Asiad, only a few upper-class individuals could afford VCRs; but when the price dropped to U.S.$900, and VCRs were allowed in as duty-free luggage of travelers, home video became extremely popular. Simultaneously, the price of a blank cassette dropped from U.S.$70 to $25 and the daily rental of a cassette from U.S.$5 to $1 and then to $0.50 (Gopal, 1986, p. 3).

The video boom hit Malaysia during the economic thrust the nation experienced in the early 1980s, when the new government initiated its "Look East" policy and sought outside investment. In 1979, 2,800 VCRs worth U.S.$2.05 million were imported; but two years later, the figures shot up to 46,000 units worth nearly U.S.$30 million. They went even higher when a 50 percent reduction in import duties for VCRs occurred in 1983. China was a relative latecomer in adopting the VCR; in 1982–1983, one scholar reported that home video was too expensive for the average Chinese family. A VCR cost over U.S.$6,000 and a cassette over U.S.$100, the latter representing a month's salary of the average worker. A few sets were smuggled in at the time, and the prediction was that the VCR eventually would cost little more than a twenty-inch color television receiver (Li Bo, 1983, p. 44). Although not providing a verifiable source, another report claimed that in 1985, 800,000 VCRs were in use in China (Ganley & Ganley, 1986, p. 35).

Other countries entered the home video generation in 1980–1981. Thailand had video clubs that charged a registration fee of U.S.$217, in addition to monthly dues. From the beginning, members illegally copied and exchanged cassettes.

THE AVAILABILITY OF VIDEO

Home video has a high penetration rate in Asia and is no longer confined just to the capital cities. In 1984, one of the authors was surprised to see a sign over the dining hall of a small hotel in Hyderabad, Pakistan, advertising "Contact Prince Video for Video Recorder";

in 1986, Einsiedel (1986) found eight video rental shops in one provincial capital of the Philippines, each averaging twenty-five rental cassettes per month for each VCR owner. An Indian researcher reported that video clubs are found even in the remote hill areas, where cassettes are carried in by bus drivers who exchange them daily (interview, Y. S. Yadava, head of research, Indian Institute of Mass Communication in Prague, Czechoslovakia, August 29, 1984).

The rural regions of India have seen a proliferation of video minitheaters (two for every movie theater) and video cafés (Sarathy, 1983, p. 53). In one of the very few comprehensive surveys of VCRs, a team of researchers found the VCR especially acceptable as an alternative to film in Indian villages. Video parlors usually begin as tea- and coffeehouses, where patrons are assessed a small fee to watch video. Once profitable, the video parlor is separated from the tea- or coffeehouse. One of the researchers said that "landlords and other prominent persons have video and in a small room of their houses, they would make video available to the public, some as an extension of their business (tea shops), some as a service to the public" (interview, Leela Rao, professor, University of Bangalore, Singapore, November 23, 1986).

If one considers only ownership of VCRs, there is some validity to Quebral's (1985, pp. 4–5) assertion that video is an urban, not a rural, medium in Asia. As she pointed out, the price of a VCR (U.S.$545) can be a poor family's annual income in a country such as the Philippines. Others have shown, however, that much sharing of VCRs occurs in poorer areas. Thus even the worst slum of Bombay has at least thirty video parlors where twenty-five to thirty people can see each of the three or four daily video showings. Outside Madras, townspeople pool their money and rent a VCR and cassette to play on the community television receiver (Rao, 1986, p. 7); and in other areas, people will rent the television set as well. Even the elite share their VCRs. In a sample of middle-class and upper-class people in Delhi and neighboring cities, Yadava (1986, p. 5) found that while 75.7 percent had television sets and only 4.5 percent owned VCRs, 36.2 percent saw video "often." A Nepalese researcher said that although most home video is concentrated in Kathmandu, some towns that receive Indian television signals will have a VCR owner who will permit other villagers to view cassettes (interview Subhadra Belbase, program director, Worldview, Singapore, November 21, 1986).

East Asia

In the more affluent countries of Asia, ownership of home video is high. Outside Japan in East Asia, Taiwan and Hong Kong have the highest concentrations. Fifteen percent of Taiwanese families own a home video for a total of 800,000. Wang (1986, p. 365) felt this penetration level placed Taiwan in the same category as the United States, Great Britain, France, Japan, and Spain. Between 1982 and 1986, during which foreign VCRs were allowed to be imported, the growth was 100 percent. The receivers are considered affordable at U.S.$500 (interview, Joe-Yang Hu, president, Empire Audio Visual Materials, Taipei, August 6, 1986). Wang (1986, p. 367) explained that although VCR prices are higher than in many countries, "they are not high enough to upset sales because of the abundance of prerecorded software." In 1985, Sony controlled 54.5 percent of Taiwan's VCR market, followed by Taiting (11.79 percent), National (10.25 percent), and Sampo (9.83 percent). Other companies included Sanyo, Zenith, and Toshiba. Sony and National maintain local assembly plants, but they and local companies such as Taiting must still depend upon Japanese parts. Nearly 61 percent of the VCRs are Betamax.

Taiwan's huge video market, estimated to be U.S.$100 million a year, depends upon 5,138 licensed videocassette rental shops and perhaps 2,000 unlicensed shops. The average shop has 6,000 cassettes in stock, and 200 new ones are issued weekly (Wang, 1986, p. 367). The director of radio and television, Dr. Sunshine Kuang, said only a portion of the videocassettes are legally registered; in fact, between 1982 and 1986, only about 5,000 were. She added that the people have "developed the bad habit of using less money to get illegal, cheap videocassettes." The public obviously gets annoyed with government authorities when they crack down on illegal cassettes, claiming they used to obtain an illegal film for NT$10 (U.S.$0.30),[1] but now must pay five times that much (interview, Sunshine Kuang, director, Department of Radio and TV Affairs, GIO, in Taipei, August 8, 1986).

Hong Kong had a rapid growth of VCRs between 1983 and 1985, from 100,000 to 750,000 units. In 1983, 16 percent of households with incomes of U.S.$2,000 or higher owned VCRs. The number of cassettes also rose accordingly, recording a 282 percent jump in the first three months of 1983 over a corresponding period in 1982. By 1985, the home video business was estimated to be HK$50–$80 million a year (U.S.$6.5 million–$10.4 million). The easy availability, at tax-free rates, of video equipment in Hong Kong is credited with spurring the growth. Video clubs have had a modest success in Hong Kong; by 1983, 35,000 VCR owners, or 30 percent of the potential market, belonged to clubs. Kam Production Studios initiated the video club idea and by 1983 had six clubs yielding a monthly revenue of U.S.$40,000. Club memberships cost U.S.$20 yearly and rentals were U.S.$2 per night. The majority of VCR owners have not joined clubs because most feature films are available on the open market in the more than sixty rental outlets operated by four groups.

Despite these figures, home video, according to Hong Kong film personnel, has had a low penetration level compared to most parts of the world. Movie director Allen Fong explained that Hong Kong

> is least hit by home video of anyplace in the world, because we are a boxed-in society, with six million people living in a small area. The people do not feel comfortable in their little homes, sitting around watching video. The young people need a place to pet each other; others to get free air conditioning, go out to movie theaters. This is good, as it stimulates the film people to do better movies (interview, Allen Fong, movie director, Hong Kong, August 11, 1986).

David Chan, vice president of Golden Communications, and Peter Lam, distribution manager of Cinema City Co. Ltd, agreed, the latter adding:

> Hong Kong has too much entertainment for people to prefer the VCR. The fantastic nightlife possible, the very small flats and young people living there with parents, encourage people to go out. With only one TV set at home, people can't relax watching video (interview, Hong Kong, August 12, 1986).

But Lam also acknowledged that as young people become more affluent, they seek a variety of entertainment, including home video. Thus, the number of rental shops has increased in recent years; beside the sixty legal ones, another fifty, called "specialty shops," admit patrons at HK$20 (U.S.$2.60) each and sell hardcore pornography (interview, Lam).

Elsewhere in East Asia, South Korea had over 200,000 VCRs and over 300 clubs in 1983–1984. Many outlets were underground, dispensing fourth- and fifth-generation copies.

Southeast Asia

Singapore and Malaysia may be the leaders in the percentage of people owning home video in this region. In late 1983, Survey Research Singapore showed that nearly four out of ten households (38 percent) had VCRs, while in 1980, the figure was 3 percent. Almost one-half of the households that owned color television receivers also owned VCRs, probably because of the relatively higher wage scales of Singaporeans and the reasonably priced sets. In fact, so many travelers from other parts of Southeast Asia and India purchase electronics products such as VCRs in Singapore that flights from Singapore to India are called "VCR flights." Malaysia and Brunei also have high VCR ownership rates—at least a million sets existed in those countries in 1983–1984. In outlying areas such as the East Malaysian states of Sabah and Sarawak and independent Brunei, where there are fewer entertainment places, the VCR is considered to be an essential item. A survey in 1984 showed that 21 percent of all adults watched video "at least once during the past seven days" (Clad, 1985). Over 2,000 video shops exist in these countries.

In the Philippines, there were an estimated 600,000 to 800,000 video owners in 1986. The prices of cassette sales and rentals are among the lowest in the world: a rental can be obtained for U.S.$0.25, and a prerecorded cassette can sell for as little as U.S.$7. When customers rent three or four titles at one time, a bonus cassette is often thrown in (Giron, 1986, p. 443). Pornography has found enough of a market in the Philippines that, since 1984, it has been profitable to produce crudely made local "boldies" that rent for U.S.$0.50 each. A full-length pornographic video can be made for U.S.$5,000, compared to the U.S.$75,000 needed for other feature films. Young people have been willing to "star" in blue videos for a fee of U.S.$750. Indonesia's video business was in a take-off stage by 1983–1984, when there were an estimated 300,000 VCRs and over 4,000 video libraries. A good-quality Japanese-manufactured VCR cost U.S.$600 to $800, and a cassette rented for U.S.$1.50 daily (Lubis, 1983, p. 54). Partly because of the prevalence of pornography in this Muslim culture, the government tried to control the videocassette industry by nationalization. Then, however, rental salesmen, usually on motorcycles, peddled door-to-door their briefcase of fifty to seventy-five titles, mostly of poor quality (Ganley & Ganley, 1986, p. 95).

Even in normally laid-back Burma, videocassettes have taken hold. One author called it one of the "most powerful vehicles" for spreading new images, adding,

> there are video rental shops throughout the city [Rangoon] and scattered around the country, offering a wide selection of movies. Indian films are popular, as are those of the spectacularly violent variety from Hong Kong and Thailand. Films from the West are also available—*Rocky* and *Rambo* have been big hits. It goes without saying that it's all pirated. According to one foreign worker here, the entire movie library of the United States Embassy in Rangoon is available for hire in the local video shops. Distribution is efficient: the 1986 World Cup football final was available here just two days later (Rothnie, 1987, p. 46).

In some cases, street cinemas are common: a screen is put up in the street and the public pays 3 kyats each to watch, the whole affair playing havoc with traffic.

Neighboring Thailand has also joined the videocassette revolution with enthusiasm. Thirty-five percent of Bangkok households already have VCRs; nationally, a high average of 20.4 percent of urban Thais possess machines (Hamid, 1987, p. 33).

South Asia

Video is difficult to escape in parts of India. There are video hotels, video buses, video restaurants, video clubs, and video theaters and halls. In 1984, when the country had 4 million television sets, there were over 500,000 VCRs—mainly in the homes of the wealthiest 5 percent of the population with incomes of U.S.$3,000 to $6,000 or above (Agrawal, 1986, p. 32; see also Agrawal, 1983, 1984). In a country that has one automobile or telephone for every 100 in the United States, the ratio of VCRs is one for every fifteen (Sarathy, 1983, p. 53). Agrawal (1986, pp. 34–35) predicted that by 1990, when the demand for consumer electronics in India will be U.S.$2.085 billion, video-related technology will account for U.S.$1.15 billion of that. He further warned that the video revolution will divert household resources from consumer items such as refrigerators, air conditioners, or scooters, to VCRs and television sets, and will lead to a large-scale drain on foreign exchange reserves and to increased external dependency.

Today, over 12,000 long-distance video buses are on Indian roads, despite a public outcry in 1983, when such a bus plunged over a cliff, killing about twenty passengers. Apparently the driver's attention had been diverted when he changed a cassette (interview, Y. S. Yadava, Prague, Czechoslovakia, August 29, 1984). In addition, at least 50,000 video libraries and 15,000 video parlors functioned in 1986. Bombay alone has over 350 video libraries, and Madras, over 250 video clubs.

Although the outlay to set up a video parlor or center for viewing is relatively high, the profits more than offset the investment. One source said that in India in 1984, it cost U.S.$2,307 for a VCR, $1,500 a month in bribery money, plus the costs of furnishing a viewing room and hiring cassettes, but that a seventy-seat parlor with three showings an evening could bring in U.S.$9,690 per month. The same source pointed out that video center entrance fees of U.S.$0.77 for a regular film and U.S.$1.90 for a pornographic one were much higher than those of cinema houses, which start at U.S.$0.15 or U.S.$0.16 in cities and less in rural areas. Those prices have been lowered to U.S.$0.14 or less (interview, Yadava). But the parlors supposedly offered more— movies that would take months, or even years, to get to cinema houses, and blue films. Describing one such parlor, *Asiaweek* of May 4, 1984 (p. 43) wrote:

In New Delhi's business district of Paharganj, a small tea shop nestles unremarkably between a grocery and a furniture store. A closer look, however, reveals more than tea tables inside. The establishment is, in fact, a plush auditorium, complete with 40 mm screen, for viewing pirated videotape films.

In Bombay and Calcutta, many high-rise buildings and apartments are connected by cable to VCR studios in their basements, which provide around-the-clock entertainment to subscribers at a minimal monthly rate. In many communities now, the video recording of special events such as festivals or weddings is available from professional sources. In fact, upper-class weddings in India are not complete unless a videocassette of the affair is made.

Spurred by the new affluence of migrant workers returning from the Middle East, Pakistan saw a video revolution of sorts in the early 1980s. Although only a minority of the population has taken to video, still, in a country that in 1983 had 80 million people (10 million households) and only 1.5 million television receivers, the 250,000 to 300,000 VCRs in existence was an impressive number. Double that number of cassettes, which sold for U.S.$20 each, were in circulation (Jabbar, 1983, p. 66). The following year, media people in Pakistan told one of the authors that the number of VCRs was 500,000 and that the daily costs of renting cassettes (Rs10, or less than U.S.$1) and VCRs (Rs40 to 50) had dropped considerably (interviews, Arif Nizami, editor, *Nawa-i-Waqt*, Lahore, Pakistan, July 9, 1984; and Syed Mumtaz Saeed, director, National Management Development Center, Karachi, Pakistan, July 18, 1984).

Sri Lanka, with a population of 15 million, has about 20,000 to 25,000 VCRs, 70 percent of them found among the middle and lower middle classes. Because of the need for viewing material, Sri Lankans early on formed clubs for cassette exchange; these quickly became commercial ventures, and by 1984 between 250 and 500 existed. The clubs carry from 250 to 4,000 cassettes, of which there are 750,000 for hire in the country. A refundable membership fee of Rs600–Rs1000 (U.S.$22–$37) is charged; the clubs make a profit on the interest gained (interview, Anura Goonsekera, director-general, Rupahavini TV, Singapore, November 20, 1986). The clubs obtain and copy cassettes from Sri Lankans traveling abroad or working in the Middle East. An article in a Colombo newspaper argued that this pirating was necessary because it allowed the middle class to see video inexpensively ("'Video piracy': The other side," 1984). Normally, a VCR costs Rs8,500 (U.S.$315) and a cassette rents for Rs15 (U.S.$0.50) daily. Video parlors, housing twenty to thirty customers, are prominent, appearing in most parts of the country that have electricity. Hindi films and materials that are banned, such as pornography, are the favorite content (interview, Goonsekera). In Bangladesh also, banned materials such as Pakistani and Hindi films are shown during regular movie nights at exclusive clubs.

THE USES OF VIDEO IN ASIA

The major use of VCRs in Asia is to entertain—more specifically, to provide materials usually unavailable or delayed through release. In Indonesia, where Chinese-language cassettes are banned, Hong Kong-produced Kung-fu movies dubbed into English have been very popular; among Malaysia's large Chinese community, materials in their ethnic languages are preferred; in Pakistan and other parts of South Asia, prohibited Hindi films are enthusiastically sought; and throughout Asia, violent and explicitly sexual films and early releases (legally or illegally obtained) of United States movies have a regular market.

In some affluent communities in India and the Philippines, among others, video has been used to record family and other special events. The video recording of weddings, for example, is a status symbol, leading to the possibility of videocassettes replacing wedding photo albums. In the Philippines, newspapers regularly carry advertisements for video coverage of special family events.

Political uses of the videocassette have been made by governments, political parties, and the underground. During the 1984 parliamentary election in India, videocassettes were used in almost all campaigns. Among them were one of Prime Minister Rajiv Gandhi's supportive speeches and one entitled *Ma* (*Mother*), which was about the late Prime Minister Indira Gandhi. Both were supplied to candidates for election campaigning purposes; the latter, circulated through 5,000 prints made at U.S.$125 each, was seen all over India (Agrawal, 1986, p. 37).

Malaysian officials used a videocassette in 1985–1986 to give the public a biased view of a riot revolving around religious issues. They obtained a police cassette of the Memali riot, edited it to screen out the actual shooting scenes, and showed it on Television Malaysia. The videocassette was preceded by a twelve-minute commentary by top governmental authorities, giving their version of the disaster (Anuar, 1986, p. 11). An anonymous source said one of the chief ministers resigned after a videocassette showing him partying strenuously was used by the opposition in an election campaign.

During the repressive Marcos regime in the Philippines, videocassettes of the assassination and funeral of Senator Benigno Aquino were copied and secretly shown in homes, where discussions ensued. In fact, a month or less after the August 1983 assassination, a pornographic video entitled, *Playboy Lovers*, was modified to include a Japanese documentary on the slaying, sandwiched between the lovemaking.

Video has been used successfully for developmental purposes, dating to at least the early 1970s, when videotaped self-awareness campaigns began in India. In one such experiment, in Saharanpur, videotapes were used to communicate with local authorities and other community groups and to increase self-awareness among villagers. The philosophy consisted of allowing villagers to use equipment, choose topics of concern in their everyday lives, and improve their self-expression, as well as provide messages to leaders higher up (Banerjee, n.d., p. 1).

Other community and religious welfare organizations in Sri Lanka, Nepal, the Philippines, the Maldives, and Malaysia have used video for development campaigns. The Sarvodaya Movement in Sri Lanka, intended to uplift the masses through self-reliance, has a separate video section. Workers take video cameras to villages to solicit people's comments on problems and their solutions. These cassettes then are shared with other villages and government leaders in cities to inform them of rural people's problems (interview, A. T. Ariyaratne, president, Lanka Jathika Sarvodaya, Singapore, Novemeber 21, 1986).

The Worldview international foundation, headquartered in Sri Lanka, is a transnational agency that has employed videocassette technology on a large scale. In recent years, it has linked the thirty-eight islands of the Maldives through a network of video parlors showing popular Hindi movies and Worldview developmental messages. Local women's groups have been encouraged to take over the parlors and use the profit for their projects (interview, Arne Fjortoft, director, Worldview International Foundation, in Karachi, Pakistan, July 7, 1984).

In Nepal, Worldview has used videocassettes to bring messages from the grass-roots level to decision makers, to train, and to take "schools and hospitals to villages" (Belbase, 1985). So far, Nepalese women have been chosen as the primary target of the Worldview project (see Appendix 4.D for a fuller description of the project). They have been taught to use video equipment to portray themselves, to discuss problems, and to create video letters on problems of nutrition, sanitation, health, deforestation, and legal rights that are sent to the authorities for possible remedy. Discussing the 1985 project, coordinator Subhadra Belbase said:

We used videocassettes to make decision makers aware of what people are saying; for educational purposes in small groups, and for training purposes. Because of video, the needs of women as they expressed them before and now will have changed a lot. We exposed them to a different world. The officials don't go to the village, so we take the village to officials (interview, Sabhadra Belbase, Singapore, November 21, 1986).

She explained that video "bulletin boards" are produced, allowing villagers to put messages on cassettes for the edification of the rest of the community. Entertainment video from outside is not needed, she said, because

the people see videos of themselves as being entertaining. They don't have other entertainment forms such as TV or movies. On the self-made videos, they can only discuss community needs, not personal ones. On a Saturday night, as many as 1,400 villagers show up to see these videos (interview, Belbase; see also Belbase, 1986, p. 5).

New Straits Times notes in Malaysia that the Trengganu Religious Department has used videocassettes to produce Muslim programs to increase religious awareness.

HOME VIDEO AUDIENCES

Because of the scarcity of reliable information on Asian home video audiences, this section highlights Taiwan, where a few surveys have been conducted. For example, a December 1982 study showed that VCR owners were generally better educated than nonowners and that 92.4 percent of the respondents watched United States and Japanese videocassettes for about an hour daily.

A series of annual VCR studies by Reng Li, a marketing survey company, revealed interesting findings in Taiwan. For example, in 1983 the VCR penetration rate of Taipei was 20.3 percent; the audience that year preferred, in this order, Japanese, United States, Chinese, and Cantonese (Hong Kong) movies and cartoons. The following year, Reng Li found the order of preference to be materials of United States, Japanese, Chinese (produced in Taiwan and Hong Kong), and Cantonese (mostly martial arts series taped from Hong Kong television) origin. Movies were the most popular, followed by drama, cartoons, and variety shows. Sixty percent of those who watched movies on video said they did not go to cinema theaters very often. Of VCR owners, 31 percent claimed they purchased their sets out of frustration with Taiwan television, and only 7.9 percent taped local television shows. The typical heavy video viewer was usually a male, 20 to 40 years old, with a higher education and a stable job. Sampling rental shop patrons in Taiwan in 1985, Reng Li found that martial arts fare from Hong Kong had risen in popularity, and that, according to country of origin, United States movies were the most frequently watched. Forty-six percent of those polled said they began their video viewing after 10 P.M.

Other Taiwan surveys in 1984, showed that the average amount of viewing time per day for home video was 2.44 hours; that 38.5 percent of VCR families watched over ten videocassettes per week, and 7.4 percent, over thirty weekly. More than two-thirds (67.1 percent) watched cassettes after television went off the air at 11 P.M. (all reported in Wang, 1986, p. 370). Another survey carried out in 1984 by Harvard Management Services, reported that 19.38 percent of video shop customers admitted to renting pornographic materials, although shop owners claimed that 80 percent did so. Nearly 43 percent of those renting blue movies were women, and of these, over 25 percent were housewives according to the *Free China Journal* of September 23, 1984 (p. 3).

In Malaysia, a survey in 1984 reported that those polled watched home video at least once weekly and that three-fourths of all viewers were Chinese. On the other hand, a 1986 survey showed that less than 20 percent of the estimated 1.3 million Malay households in the country had VCRs. The Chinese prefer video originating in Hong Kong or Taiwan because the television services are predominantly in the national language, Bahasa Malaysia, which many older Chinese do not understand. There has been some speculation that the market is ready for the production of Indonesian and Malay videos, according to the national daily, *New Straits Times*, of October 15, 1986.

Thais turned more towards video, television, and magazines, and less to newspapers and radio between 1985 and 1986, according to a 1986 Deemar Media Index. Of respondents 12 years of age or older, video viewership rose by 25 percent, and television viewership by 12 percent. Newspaper readership dropped by 4 percent, and radio listenership by 10 percent. Rural Thailand, where there are 32.8 million people, showed much higher increases; video viewership shot up 62 percent, TV by 14 percent, and magazine reading by 9 to 25 percent. Newspaper readership in upcountry Thailand had a 10 percent drop; radio, 12 percent; and cinema attendance, 5 percent (*AMCB*, March-April 1987, p. 4).

An all-India survey on home video, still in the writing stage, has been carried out by the Indian Social Research Organization in the North, South, East, West, and Northeast regions. Preliminary findings on the South indicate video, especially in parlors, is growing rapidly. Movies of all types are popular; and high-action, violent, Dracula-type horror, mystery, and pornographic films are the most favored foreign ones. The most popular Hindi films carry religious and social themes. The researcher responsible for that region explained:

> Violence and pornography are not culture specific, but social-religious themes are. People want to see their own country's films on these themes. The differences in viewing are that high-action thriller films have young boys as an audience, while social-religious themes are mainly family shared. Social-religious themes would be anything promoting faith, belief, the triumph of good over evil, and the importance of family—in other words, the typical Indian film (interview Leela Rao, Singapore, November 23, 1986).

Rao also said that regional language films were preferred video fare in most states, but that the language preference varies from state to state. In Tamil Nadu, in an effort to protect the local industry, Tamil films are sought; while in Andhra Pradesh, a mixture of films in the national and local languages is used—Hindi in video parlors and Urdu in homes (interview, Rao).

SOURCES OF VIDEOCASSETTES

Because home video is not a formal medium that is controlled or monitored by the government, and because it uses highly accessible and portable equipment and software, its sources are often difficult to locate. Listening to film and video personnel discuss the ways in which VCRs and cassettes enter Asia can be a lesson in ingenious marketing (and thievery). In Taiwan, videocassettes of the previous night's television shows from Japan are available through an informal network; someone in Tokyo tapes off the TV screen and then arranges for tourists or students to carry the cassettes on a night flight to Taipei, where a pirate makes copies throughout the night. In fact, about twenty-five cassettes arrive in Taiwan by this route every week (Wang, 1986, p. 368). In India, videocassette recorders arrive from Dubai and are unloaded nightly from *dhows* (small boats) into caves and inlets near Bombay. In addition, Madras and Bombay merchants hire couriers, and provide them with passports and airline tickets to travel to other countries to shop for video equipment (Ganley & Ganley, 1986, p. 34).

Agrawal (1986, p. 32) could have been describing the Asian scene generally when he gave three sources of VCRs in India: from the original manufacturing country; from video kits imported by multinational corporations for assembly in India; and from individual importation, both legal and illegal. Concerning the latter, tourists, relatives, and friends traveling abroad, as well as government officials and migrant workers, enter customs docks throughout Asia loaded down with electronic paraphernalia, especially VCRs.

Cassettes have similar sources that might be categorized as legitimate importation, local production and distribution, and pirating and smuggling.[2]

Local Distribution and Production

In Taiwan, for example, the largest distributor of imported videocassettes is Dynasty Video Enterprises, which supplies 1,800 rental shops with a yearly average of 450 titles. Dynasty imports mainly from the United States, Hong Kong, South Korea, and countries of Europe, maintaining contacts with individual production companies. About a dozen videocassette distributing companies exist in Taiwan, some of which specialize in the popular Chinese movies produced in Taiwan and Hong Kong. The large number of video rental shops in Taiwan (5,173 registered as of July 1986; 1,500 to 3,000 not registered) requires a steady supply of cassettes. Between 1982 and mid-1986, of 5,566 registered videocassettes in the country, 4,387 were foreign and local films, including features made in Taiwan; 391 were television programs, and 788, local video features (interview, Sunshine Kuang, Taipei, August 8, 1986).

The number of local video production companies in Taiwan is hard to determine; of the 257 listed, only 25 are active. The president of empire Audio Visual Materials gave the number of local production and distribution firms as 35. Among them, Empire and Century are the major distributors of the popular videocassettes of Hong Kong television. Empire imports 2,500 Channel ATV programs yearly, and makes 4,000 to 10,000 copies of each for Taiwan video shops. The more popular shows have even more copies. Century brings in a similar number of videotaped Hong Kong television shows. Hong Kong television videocassettes, which cost NT$50 (U.S.$1.50) each, are popular in Taiwan, especially after local channels go off the air at 11 or 11:30 P.M.

Local video production houses have developed in parts of Asia. In Taiwan, a television personality began to produce videotaped materials (mostly soap operas) in the late 1970s; by 1984, when the number of companies peaked at forty, local production seemed on the verge of becoming big business. But it did not flourish for long because television stations refused to buy the cassettes unless they were released first through them, and rental shops did not have confidence in local production houses because of the haste with which most developed. As a result, many production companies closed before they made their first videocassette; by September 1985, only five active producers remained. The largest, Yung Chang, released eighty cassettes in 1985, although claiming 200 (Wang, 1986, p. 369).

Empire, connected with CTS Television in Taiwan, does video for military and educational purposes and reproduces 500 CTS television programs yearly, of which 300 are exported, mainly to Malaysia and Singapore. After a CTS show is on the air, Empire has permission to reproduce it anytime it wishes (interview Joe-Yang Hu, Taipei, August 6, 1986).

The government-backed Central Motion Picture Company (CMPC) has its own video production house linked with a subdivision of Thirty-One Audio Visual Company. CMPC copies and distributes on videocassettes feature-length movies made in Taiwan since 1949, original stories, and United States and European quality movies. Local Mandarin movies must be in circulation at least six years before CMPC is allowed to copy them. Yearly, CMPC releases twenty Mandarin films, eight new serialized local productions, and six United States and European movies (interview, Benny C. P. Chao, director of production, CMPC, Taipei, August 5, 1986). Hoping to fight piracy, CMPC has signed a contract with 1,600 video rental shops to distribute these cassettes, each shop providing a NT$10,000 (U.S.$300) guarantee.

Elsewhere, HK-TVB International was formed in 1975 to distribute Hong Kong Channel TVB programs to the international market. Less than a decade later, HK-TVB's programs were sold to broadcasters in fourteen countries, and the organization had home video operations in nineteen. HK-TVB licenses its program rights to operators in as many countries as possible, using a "low-price policy to encourage trade entry so that the licensees, for their own interest, would take action against the pirates" (Cheung, 1984, p. 17). Thus, HK-TVB grants to its licensees the right to duplicate its programs in their respective territories and supplies the masters immediately after their first broadcast in Hong Kong.

In 1985, Indian film producers began to combat the "video menace" by selling video rights to their work and to put out videocassettes of their own. Similarly, Bangladesh Radio and Television, among a few other Asian Broadcasters, offsets piracy and competes with home video by producing its own audio and video music cassettes for sale to the public. Broadcast authorities also hope to drive out low-quality videocassettes, especially those containing content not relevant to Bangladesh (Mustafa, 1987, p. 18).

Pirating

The pirating of videocassettes is an immense business and probably the largest source of video materials in most parts of Asia.[3] For years, the pirates gripped the markets in Southeast Asia and Taiwan so firmly that some U.S. and European distributors refused to release their films there. One industry source was quoted as saying:

> It's hard to blame them. When they look at Southeast Asia, as well as Taiwan and Hongkong, they just see a giant skull and crossbones. Why should they operate here, when they know they're going to be robbed? ("Video pirate scourge," 1982, p. 46).

Everywhere in Asia, cassette piracy is widespread. Illegal Indian films appear on the streets of Pakistan in a couple of days via the Dubai connection; pirated films from India, Hong Kong, Singapore, or London widely circulate in Sri Lanka; and pornography is available, usually under the counter, in all countries.

Of course, piracy is prevalent in other parts of the world; the difference is that in Asia, the pirates have operated openly without much governmental interference. Thus, on the streets of Manila, the many video shops display hundreds of cassettes, many of which have been recorded from cable and network television in the United States, and keep detailed reference lists to guide buyers in their choices. One source wrote that

> With commercials edited out, a whole season's run of "*Dallas*" or any other American TV hit can be compiled on just a few cassettes, each of which is reproduced thousands of times by the pirate. The cost is minimal, the profit is enormous, and the risk is zero ("Video pirate scourge," 1982, p. 46).

In Taiwan, considered to be the mecca for pirates, it is estimated that nearly all 8,000 video shops sell and rent pirated materials. The general manager of the largest importer of United States movies said it has to be that way, claiming legal shops, which must pay royalties on cassettes, cannot compete with pirate-oriented video shops, and therefore, join them (interview, Frank S. L. Fan, Metro-Goldwyn-Mayer of China, Taipei, August 5, 1986). Another writer explained that piracy is difficult to tackle in Taiwan because (1) Chinese culture does not favor the concept of copyright; under the Confucian tradition, the great literary works have been passed on freely; (2) the laws are outmoded; and (3) structural deficiencies exist in that not enough antipiracy staffing is available (Wang, 1986, p. 374).

The openness with which some pirates work is exemplified in Malaysia and the Philippines, where they are even organized. Nearly 300 operators in Malaysia combined into a registered group that, according to one copyright enforcer, "has tied up distribution systems so effectively that legitimate companies don't have a chance of getting a foothold" ("Video pirate scourge," 1982, p. 46). The Philippine pirates were even bolder; they organized into a syndicate, whose purpose has been described as "bleeding the movie industry to death." The syndicate waged an aggressive campaign against the Video Regulatory Board, rallying at its headquarters and mounting media campaigns. Some public support has been given the pirates because of the assurance of lower consumer prices and because of the syndicate's use of nationalistic concerns, claiming that video regulatory laws favor foreigners (Giron, 1986, p. 443). In 1982, when nine members of the Philippine Video Association were raided by the police, the pirates sued the raid instigators.

In India, where a 1984 copyright amendment had some curbing impact, the pirates still crop up, and sometimes in elaborate settings. Basu (1984, p. 40) said that a video library raided after the new law

> was almost an arcade: a library, a video parlour and a "transferring unit" in an adjoining residential flat. Nine video cassette recorders (VCRs) were busy taping when the raiding party got there.

Although the law is tough, pirates are not often apprehended because they provide the police with a large part of their *hafta* (bribes). Even though the police, linked up with the Film Producers' Welfare organization (FPWO), raided fifty to sixty libraries and minitheaters in Bombay during a two-week period in 1984, their marks overall were not good. FPWO Head Vikas Moham, describing the problem of getting police to act, said:

> First, we would file a complaint with a police station, and the officer would procrastinate, call for files, keep us sitting. So we went to [the] commissioner, who promptly sent out a circular to all police stations in the city telling the officers to give us all help (Quoted in Basu, 1984, p. 40).

After pirated or pornographic cassettes were seized by police, they "miraculously" were replaced during the same night by perfectly legal ones or blank cassettes. FPWO plugged this leak by requiring squares of paper on all impounded cassettes with the signatures of the FPWO head, the police officer, and the culprit.

One source of pirated materials is the master print itself. Self-serving employees in film laboratories, as well as some legitimate producers and distributors, have made films available to pirates simultaneously with their official release. One film man in Taipei said he saw no other way for these films to get onto the market so quickly, adding:

> The pirates must have good sources; they must be getting their videos from the United States producers and distributors. When a picture is released in the United States the master comes here in a week, and the next day, 2,000 copies will come out (interview, Danny Tchill, owner, ESC Inc., Taipei, August 5, 1986).

Indian officials have warned film producers to watch their spools in theaters' projection rooms, and to send their prints for screening abroad "under tight security." Nevertheless, the pirates seem to break into the security, as reported by Basu (1984, p. 40):

> Normally delivery of spools to the overseas market has to be effected 15 days before the film's release. Dev Anand thought he could bug the bugs; he sent the spools of *Anand aur Anand* to London on Wednesday, under the impression that not the wiliest of pirates could get hold of the print in time to beat the film's release in India on Friday. On Thursday evening, he was informed that the pirated cassettes of his movie had arrived in the Bombay market.

In the Philippines, sources of pirated video include film laboratories and unscrupulous buyers, and some film producers who resort to double selling. In addition, video versions of Filipino films are released in the United States concurrently with their Philippine exhibition, and almost immediately, pirated copies find their way back to Manila (Calderon, 1986). In many nations, pirates copy films from the theater screens. Taiwanese pirates, for example, take their cameras to Saturday midnight previews and record movies, complete with the sounds of audience reaction and views of the backs of theatergoers' heads (interview, Jen Wan, film director, Taipei, August 7, 1986). Still other piracy techniques include copying from foreign and domestic cable and network television screens (sometimes, but not always, editing out commercials), and from legally released films, either done locally or brought in from abroad.

GOVERNANCE

Home video, and especially the aforementioned piracy associated with it, is very definitely on the minds of government officials in Asia. They have been concerned for a number of years. For example, in 1982 the Malaysian information minister hit out at home video and within a couple of years, that country tried to remedy the problem by developing a third television channel. President Zia of Pakistan, addressing film people in 1984, pleaded for their help in coming up with solutions. Other national leaders, such as those of Taiwan, introduced home video legislation back in the late 1970s. Although admitting to a sort of helplessness in controlling home video, the authorities nevertheless make an effort with admonitions, legislation, and competing business practices.

Often, they were encouraged by actions taken by the United States government, in concert with the private sector. In 1984, the United States International Trade Commission (ITC) released a report in which it claimed that in 1982 alone, the effects of all foreign-product counterfeiting were a minimum financial loss of U.S.$6 billion to $8 billion and over 131,000 jobs. Taiwan was the leading source of counterfeiting, followed by Hong Kong, Indonesia, Singapore, South Korea, and the Philippines. To attack the problem, the United States government has worked on at least two fronts: internationally, by soliciting the help of European countries and Japan in anticounterfeiting efforts; and domestically, by altering trade preferences to favor countries that practice copyright safeguards and by drastically increasing penalites for trafficking in pirated goods (Manning, 1984, pp. 62–62). Of the leading offenders, Taiwan was more willing to move against pirates than was Singapore (Kaye, 1984, pp. 82–83). But by 1986 both countries, as well as others in Asia, had gotten the message. For example, South Korea planned to revamp its 1957 copyright law as a result of the pressure.

Outside pressure has also been applied by the International Federation of Phonogram and Videogram Producers (IFPVP) and the Motion Picture Association of America (MPAA), both of which keep representative in the region. For years, the IFPVP has tried to persuade Asian governments to act on copyright protection. Only recently has the organization been somewhat successful, partly because its meetings no longer just assess damages, but work out the implications of new copyright laws. The president of IFPVP, Nesuhi Ertegun, in late 1986, said that for years, Asian governments lacked sincerity on the issue of piracy, which "endangered the future potential of foreign business interests in this area," but that the situation had improved

(Leo, 1986, p. 15). The MPAA claimed that its members lose at least U.S.$11 million yearly because of illegal videotapes in Singapore alone (Holloway, 1986, p. 58).

As alluded to previously, home video poses a particular headache for those responsible for its monitoring and control—it is elusive, being predominantly an in-home system, and it is very fast developing. In fact, home video has come onto the scene with such rapidity that the quick amendment of antiquated copyright, obscenity, and pornography laws has not been possible.

Once over their initial surprise at the ascendancy of home video, governments did amend existing copyright and obscenity laws, as well as institute new ones. Additionally, they tried other controls such as registration of VCRs and cassettes, increased import duties on home video materials, more frequent police raids on video stores and parlors, and development of regulatory boards.

East Asia

Of all Asian countries and territories, perhaps Hong Kong has been the most successful in the governance of home video. Once a major producer of pirated videocassettes, Hong Kong changed its image considerably after 1978, when forty-two full-time police were employed to deal solely with copyright infringements, and cooperation was increased between the Hong Kong Customs and Excise Department and the local branch of the IFPVP. In 1984, a government committee attacked the sale of pornographic video by granting authorities the right to censor cassettes before their sale. The action came after the legitimate video industry asked for protection from fly-by-night producers and distributors who marketed pornography. Copyright infringers can have their premises searched, have all materials seized, and be fined HK$1,000 (U.S.$130) and jailed for a year. There is no ceiling on the number of infringing copies to be charged.

Realizing that piracy cannot be stopped by force alone, Hong Kong TVB International has gone beyond helping police with raids, to devising an attractive marketing concept that entices pirates to become sublicensees. The head of HK-TVB explained:

A market vacuum created by the arrest of one pirate will immediately be filled up by a second pirate. We must therefore couple these actions with strong marketing back-ups such as quality of duplication, attractive packaging, extensive distribution, etc. so as to *replace* piracy, albeit a painful, slow and very often expensive process (Cheung, 1984, p. 18).

He provided an example of one such conversion made by HK-TVB in Malaysia:

> In 1979, a Mr Lee was running a rubber factory in Ipoh. Because of his frequent visits to Hong Kong, he soon realised the quality and potential of TVB programmes but, instead of approaching us to try and obtain a licence..., he requested a friend in Hong Kong to record all the TVB programmes and mail the cassettes to him. Then in Ipoh he set up a small coffee shop and started to play these cassettes in the shop and charge customers a bit higher on the drinks. Business flourished and while VCR machines increased in number, he started to duplicate his cassettes and rented them out. He made a small fortune. Others followed him and several pirate operations started.
>
> We concluded our first Malaysian video licence in 1980 and immediately our licensee approached Mr Lee to talk him into becoming a sub-licensee. Our selling point was simple but effective: with an exclusive licence Mr Lee could stamp out his competitors in Ipoh and therefore enjoy a monopoly and make more money. Mr Lee readily accepted the licence since the fee was very low and proceeded to raid his competitors. He soon realised, however, that copyright enforcement was no easy task and moreover, the market was growing so quickly that his own video shop could not handle the business effectively. So he changed his approach and push-pulled the pirates into obtaining sub-licences from him to continue with their operations. Today Ipoh has about 10,000 VCRs and Mr lee has 45 licenced dealers, each paying him an annual fee of 10,000 Malaysian dollars, and piracy is kept to a very low level. On the other hand, our royalty fees receivable from Malaysia as a whole has [*sic*] increased more than 20 times within three years (Cheung, 1984, p. 18).

In Taiwan, the solutions to the video dilemma have not proceeded to this stage. Government officials and electronic media personnel there express frustration when discussing home video control. For instance, Dr. Chang King-Yuh, then director-general of the Government Information Office, said:

> A problem here is that the government does not change as fast as the technology. Videotapes are a serious problem. They don't respect copyright. We have thousands of stores with videotapes and it is hard to police them. Restaurants and other entertainment places are showing video without registration. How do we enforce the law? We are surrounded by seas and smuggling of motion pictures by air and ship is prevalent. Most are X-rated offering sex and they have not been reviewed by the board. We raid some of these but are not very successful (interview, Chang King-Yuh, Taipei, August 6, 1986).

However, some progress was made when Taiwanese laws were toughened between 1981 and 1986 to deal with intellectual property rights. In 1981 customs made exporters show their authorization to use foreign trademarks; in 1983 and 1985 penalties were raised in the trademark law, and in 1985–1986 work was completed on amending copyright regulations.

The Enforcement Rules of the Broadcasting and Television Law, issued in 1983, specifically defined videotapes in Chapter 5, "Video-Tape Control," and provided application procedures in gaining permission to issue tapes and import-export requirements. Under the law, a certificate is needed before distributing or broadcasting videotapes, all of which have to carry identifying stickers. Further, conditions under which tapes can be broadcast were set: if only one audiovisual machine is used for broadcasting a videotaped program, the broadcast can be made at home; otherwise, the location of the broadcasting must be a site set up according to law by a film-screening establishment and the videotape being screened must have a certificate (see portions of the law in Appendix 4.A).

The 1985 revision of the copyright law made the pirating of videocassettes not just an administrative but a criminal offense, leading to a possible five years in prison, a fine of U.S.$7,500 and confiscation of the pirated goods. Later that year, the Supreme Court of Taiwan ruled that the renting of pirated tapes also constituted a criminal act (Wang, 1986, p. 375). When the *Copyright Law of the Republic of China* appeared in 1986, it included videotapes among works protected and granted reciprocal protection to foreign nationals whose works were first published in Taiwan. The punishment for reproducing copyrighted materials was imprisonment for at least six months and not more than three years, and a fine not to exceed NT$30,000 (U.S.$900). The illegal selling or leasing of pirated materials carried a prison term not to exceed two years and fines under NT$20,000 (U.S.$600).

The director of Radio and Television Affairs of Taiwan, Dr. Sunshine Kuang, believes the broadcast law needs more revision, as it does not deal effectively with video. She said the law specifies that video cannot be shown in public places except in theaters; but, according to her, this is ridiculous because theaters are not appropriate for video showing.

Kuang classified the antipiracy legislation in the copyright amendments as satisfactory, but she said their enforcement needs much more work, adding:

> There are only nine persons in my video section, so it is hard to enforce the law. We visit illegal shops, according to reports from the

public. We check tapes and confiscate, but we have no power of arrest. If a tape is pornographic, we pass it on to the police to make an arrest. Sometimes, distributors or the Union of Videogram Distributors will report illegal shops to us (interview, Sunshine Kuang, Taipei, August 8, 1986).

The Office of Radio and Television Affairs also screens all legal videocassettes before their distribution. About 5 to 10 percent of them are banned, usually because of their sexual or violent content. The office has a censoring pool of fifty to sixty individuals, from whom it chooses one to review each videocassette. The office, according to its director, is in a no-win situation concerning reviewing, as "the conservatives say we let sex into the country and the liberals say we are too strict" (interview, Kuang).

Other groups have been involved in eliminating video piracy in Taiwan. The Government Information Office's Department of Motion Pictures thinks clearing out the pirates is its biggest concern (interview, Chiang Tsou-Ming, director, and Liu Shou-Chi, section head, Department of Motion Pictures, GIO, Taipei, August 5 1986), while the Motion Picture Association of Taiwan (MPAT) has developed an antipiracy committee that systematically investigates video piracy and reports incidents to the police. The head of MPAT said the government looks favorably upon information gathered by his committee; but because of the low priority of video raids in the overall national scheme, he has to continually prod authorities to take action. The committee is better able to find pirates of locally made films because of the registration numbers they receive when produced. With foreign-made movies, MPAT immediately registers them locally to stop pirated versions later (Interview, Chou Ling-Kong, director, Fee Tang Motion Picture Co., Taipei, August 6, 1986). With government assistance, an association of audio- and videocassette producers, distributors, and retailers was established in the mid-1970s to exercise self-discipline. The government requires that all applicants for rental shop licenses first join the association.

Yet these legislative changes and the linking up of organizations for control have not changed the situation very appreciably. The government has confiscated over 20,000 pirated cassetes but is limited in expanding its work because of the large number of rental shops to police and the small staff to do it with (Wang, 1986, p. 375). As one distributor noted, "The problem is so overwhelming that we feel impotent. The law is there, the regulations are there, but piracy still thrives" (interview, Tchii).

China has taken some action to regulate its chaotic home video market. Without an effective copyright law, the reasons usually were for other than protecting intellectual property rights. One main concern has been with "cultural pollution," the bringing in of outside values and ideologies not conducive to Chinese culture. For example, on March 13, 1982, the State Council instituted an order banning the importation, reproduction, sale or showing of "decadent and indecent" records, cassettes or videotapes (Li Bo, 1983, p. 44). Laws affecting video were beefed up in 1986, when the State Council and the Communist Party Central Committee declared that the purpose of importing is to meet the needs of the public for foreign cultural programs. A November 1986 directive spelled out the terms under which videocassettes can be sold, aimed at stemming the flow from private sources. Officially, videocassettes must be distributed through the China International Television Corporation (ITV), which selects and distributes industry-approved titles to provincial "broadcast departments," which, in turn, release them to similar operations in the towns. These commercial outlets are privately operated quasi-governmental agencies functioning in partnership with ITV. By keeping all duplication and distribution within the ITV structure, the new directive allows the government to offer copyright protection to licensors whose registration documents are submitted to the ministry. Private entrepreneurial video was made illegal by the rules. In an article by Mark Silverman, *Variety* (March 5, 1986, p. 1) commented that

> Rental of videocassettes would be cleared for so-called "private showing"—instances of small multifamily units viewing collectively, but without separate admission charges.... Since the idea of individual ownership of property is contrary to the Communist ideology, this is a rather novel step.

The antipiracy policy in the directive warned that illegal video is of "poor quality and unhealthy," and gave reasons why its impact will be resisted:

> So as to enforce and strengthen the control on the audio and visual products in the market, protecting the majority's interests, and in order to benefit the spiritual foundation of socialism, the authorities have decided to restrain and give great blows to illegal product duplicating and merchandising.

Describing the consequences of the sale or possession of illegal videocassettes, the directive added:

> The unlawfully duplicated videocassettes found in the market would be confiscated, as well as the sales income. If all copies have been

sold out, the fine will amount to three to five times the amount of the purchase corresponding to the seriousness of the violation. Stock of the illegal duplicates will be confiscated. Besides, he will be fined and restitution be made to the (video) publishing units according to the economic loss resulting from the illegal reproduction. All the confiscated videocassettes should be cleaned [erased] by the TV broadcasting department (in *Variety*, March 5, 1986, p. 1).

Penalties also applied to the illegal viewing of videocassettes. As if to highlight their point, in 1986 the authorities expelled a famous Chinese writer from the Communist Party for watching pornographic videocassettes.

Southeast Asia

Countries in this region, well known for their video piracy, began to clean up their acts in the mid-1980s, amid increasing impatience in Washington. Singapore moved most swiftly. In 1986, the government of Lee Kuan Yew came up with a new law, providing fines of U.S.$4,650 to U.S.$46,500 and jail terms of two to five years, for the illegal reproduction of intellectual materials. Before this law, pirates were willing to risk fines of a maximum U.S.$984 and a year in jail if convicted, which were small punishments in comparison to the 35 million pirated cassettes worth U.S.$60 million made in Singapore yearly.

Deputy Prime Minister Goh Chok Tong, reacting to complaints of government inaction, said that Singapore copyright was based on 75-year-old British laws; he also claimed the government had been acting, seizing over 49,000 illegal tapes in eighty raids during the first seven months of 1985 (AP dispatch, December 8, 1985). Crackdowns on the circulation of pornographic videos had been carried out since the early 1980s.

In 1983–1984, after a year's deliberation, the government instituted systems of licensing and censorship to govern video. Four types of licenses were put into effect: to import pre-recorded tapes, to record and distribute tapes, to exhibit tapes, and to distribute tapes. The fourth type was meant for libraries, while the third for hotels providing in-house movies. Legitimate video traders had to pay a deposit of S$20,000 (U.S.$9,400) and S$1,200 (U.S.$564) for an annual license, and have a paid-up capital of S$200,000 (U.S.$94,000). Any act of piracy resulted in forfeiture of the fee and other severe penalties, including the loss of rights in perpetuity for public housing. *Variety* (May 9, 1984) reported that the stringent regulations actually helped piracy, rationalizing that since it took a year for a tape to be released because of the slow censorship process, the pirates flourished in the interim. A 1985 system was designed to cut the time needed for censorship by two-thirds by having censors look only at a master copy and accept the word of the libraries that all copies were indentical.

Neighboring Malaysia's copyright law has also been slightly changed from its 1969 form, and an even tougher draft is in the works. It supposedly unequivocally covers videotapes and raises the max-imum fine two-and-a-half times over the present U.S.$38,536. The 1969 law was severe in itself, but it remained unenforced for twelve years and its main provisions were not upheld until 1985. In that year, the Supreme Court declared that all foreign works published in Malaysia within thirty days of first publication overseas would enjoy copyright protection (see Menon, 1984).

Indonesia updated its copyright law in 1982. Before that, the act was based on old Dutch regulations and did not cover protection of video or foreign works. Thailand, one of the first Southeast Asian countries to protect copyrighted work, has been lax in enforcing the law, partly because of its complexity and the difficulty of obtaining documentation to justify prosecution. Thus, the piracy of videocasset-tes is extensive, and is practiced by a chain of financially secure and well-organized commerical enterprises ("No hiding place," 1986, pp. 58–59). When the United States threatened Thailand with trade restric-tions in 1987, if something was not done to protect copyright, a furor was raised in Parliament. In July 1987, an amendment to the 1978 copyright law was approved, granting the United States the same protection Thailand grants signatories to the Berne Convention. With the strengthening of copyright regulations in these countries, video, film, and other local entertainment industries hoped to enhance their markets.

Indonesia did more than just update its copyright laws. With Presidential Decision 13 of 1983, labelled one of the "most strenuous and sweeping efforts of any country to control videocassettes," the government nationalized the video rental industry. Only three state-owned companies were authorized to import, reproduce, or distribute any type of video. Furthermore, various documents on the importa-tion, ownership, censorship, and intended use must be filed before these companies can take any action. Only master copies of cassettes are allowed in, and reproduction has to be done locally (Ganley & Ganley, 1986, p. 144). The government had made decisions on video from at least 1979, when the attorney general, whose office is in charge of VCRs, declared that all cassettes imported into Indonesia had to be carefully scrutinized and any violation of the nation's stringent obscenity laws would be dealt with severely. In 1985, the Indonesian National Film Council Published *Guidance on Video Recordings in Indonesia*, listing seven decrees written since 1983 and carried out by the Ministry of Information. The decrees specified conditions on "video recording companies," "license for producing Indonesian video recording domestically and abroad," "video recording of joint production between Indonesian production company and foreign production company," "provisions in the field of video recording importance," "video recording multiplication," "video recording distribution procedure," and "video recording broadcasting and showing."

The Philippines, in an effort to deal with the U.S.$30-million-a-year business pirates had built there between 1978 and 1986, created the Videogram Regulatory Board (VRB). Presidential Decree 1987, which established the board, was a reaction to the losses the government faced in tax revenues, and the film industry, in box office billings. The twenty-member organization, appointed by former president Ferdinand Marcos, is empowered to act against video that is

> objectionable for being immoral, indecent, libelous, contrary to good law or good customs, or injurious to the prestige of the Republic of the Philippines or its people, or with a dangerous tendency to encourage or fan hatred, the commission of violence or of a wrong or crime.

More specifically, the act defined censorious content as that which incites insurrection, subversion, or rebellion; undermines public faith in government; glorifies crime, excessive violence, or sex; is libelous; abets traffic in drugs; or commits contempt of court. The board is authorized to rate videocassettes, close illegal establishments, review tapes for censorship purposes, and levy and assess taxes. Traders in video are required to register with VRB, and violators of any board regulations face potential three-month to one-year jail terms and fines of P50,000–P100,000 (U.S.$2,500–$5,000). Provincial governments can collect a tax of 30 percent on the sale or rental of a videocassette (see P. D. 1987 in Appendix 4.B).

In 1986, the VRB and its provisions were strengthened. With a staff of fifty-nine people and the job of overseeing more than 5,000 video establishments, the VRB had been ineffective, prompting one critic to write that without police powers and with just a small staff, it had tried to collect the 30 percent tax (Velarde, 1986, p. 15). The new guidelines provide for 1,000 VRB representatives with powers to arrest, sequester or close down illegal operators (interview, Lamberto Avellana, director, Documentary Inc., Quezon City, Philippines, August 16,

1986), and require the affixing of VRB seals on cassettes. Also, all operators must obtain a VRB rating for cassettes; the ratings after May 1986 were "general patronage," "parental guidance," and "restricted" (Pacheco, 1986; Calderon, 1986). Previously, videogram establishments could duplicate tapes without permission from the owners, provided they put up a cash bond. Under the new rules, this is not permitted. Generally, the new law limits the number of duplication and rental outfits, raises taxes, provides more censorship and classification powers, increases the strictness on piracy and pornography, and requires the VRB seal (see Siguion-Reyna, 1986).

In mid-1986, newspapers lent support to a campaign to clean up video in their advertisements and columns. In one such column, the head of VRB, Eduardo Sazon, gave his solutions to the video problem as (1) conviction of video pirates, (2) economic cooperation between the movie and video industries, (3) professionalization of the video industry, (4) self-regulation among the video establishments, and (5) development of proper linkages to other government agencies and the military (Rodriguez, 1986, p. 11). An interview with Sazon can be found in Appendix 4.C.

Some critics remain skeptical, claiming that although the government has been making rules for years, video pirates still thrive. As early as 1981, the Board of Censors required public exhibitors of video—i.e., hotels, pubs, and restaurants—to register with the authorities and to obtain written permits for all cassettes they exhibited (Calderon, 1986). Other writers predicted that the new rules will mean an increase of 25 to 50 percent in rental costs to customers, as retailers pass on the annual P5 (U.S.$0.25) per tape excise tax and 30 percent amusement tax they must pay (Pacheco, 1986).

South Asia

South Asia has not advanced as quickly as other parts of Asia in its strengthening of legislation. The result is that Sri Lanka has done virtually nothing, while Pakistan and Nepal have instituted high import duties on VCRs. In Nepal, the customs duty is 100 percent; and the authorities, after otherwise trying to control video, now ask only that the tax on the showing of video in parlors be paid (interview, Belbase).

Pakistan, in 1983, doubled the import duty on VCRs. But a year later, a Pakistani media official said the duty was to be lowered from 100 to 85 percent as a "way of the government saying it cannot control the import of video and it will do away with import duty altogether soon" (interview, Abbas Muzaffar, director, Pakistan Information Service Academy, Lahore, Pakistan, July 12, 1984). Others also pointed out the frustrating position officials found themselves in (interview, Arif Nizami, in Lahore, Pakistan, July 9, 1984). Guidelines on pre-recorded cassettes did not exist, except that Indian films on cassettes were seized because of a 1965 ban. One writer said that arrests were made only when there was a "falling out" between law enforcers and lawbreakers, or when the violators were so blantant about their opera-tions that the police were compelled to act out of embarrassment (Jabbar, 1983, p. 65). The same author described the confusion around the policing of video as a "Keystone cops farce," adding:

> No week passes without the police in some area of Karachi raiding premises where cassette shows of Indian films are being organised on a commercial basis by plucky local "videoentrepreneurs" for eager viewers who pay anything upwards of ten Pakistani rupees (about U.S.$0.80) for two hours of loud and colouful entertainment. Some-times the evening papers publish pictures of the organisers being arrested, doing their best to look ashamed, and failing in the attempt. For it is public knowledge that virtually every house which has a VCR machine (and in the major cities and towns every second home in every high-income and every upper-middle-class neighbourhood/ area probably has a VCR set) is bound to be screening an Indian movie at one time or another without suffering any police raid (Jabbar, 1983, p. 65).

In Sri Lanka, the newness of video and, in fact, television means that there is not much of a regulatory framework. The laws of the Sri Lanka Rupahavini Corporation (Television) list, under functions and duties of the corporation, "exercise of supervision and control over the use of video cassettes and the production of programme material on such cassettes for export." However, early on, the chairman of Rupa-havini did not know how to implement this power; and when he sought advice from the Ministry of State in charge of broadcasting, the latter disclaimed responsibility (Kurukulasuriya, 1983, p. 69). More recently, the director-general of broadcasting said that a committee, appointed by the Ministry of Justice, is looking into "ways and means of controlling video." His fear is that the film lobby will be successful in having video banned, at which time it will go underground. "It is more sensible to tax and control video, but not ban it," he said (inter-view, Anura Goonsekera, Singapore, November 20, 1986). But in December 1986 Sri Lanka was still an open market for the cavalier video dealers, who have very little to fear, with the lack of laws and the relative inattention of the government (see another view of the problem in Appendix 4.E).

India, mainly because of pressure from the large film industry, has taken action since 1983–1984. According to Agrawal (1986, p. 36), the government imposed a fee in November 1983 for the use of VCRs, owners being required to have a domestic or commercial license. In addition, several state governments have imposed similar fees on VCRs used for public viewing.

There have been other changes. In 1984 the government enacted a number of bills and amendments regarding video, one of the most stringent being the Copyright (Amendment) Act, passed on August 27, 1984, which provided for severe punishment for video copyright offenders. The action came when it did, according to some political pundits, because 1984 was an election year and fund collection time for political parties. A new section requires video producers to display on the cassette and its outer cover, the names and addresses of the producer and copyright owner. Another section specifies that for a second-time video offender, the punishment will be imprisonment for not less than a year, but extendable to three years, and a fine of not less than Rs1 lakh (U.S.$7,000), extendable to Rs2 lakh (U.S.$14,000). Section 64 has been amended to empower "any police officer, not below the rank of sub-inspector," to seize, without warrant, all copied cassettes, as well as the master, and present them before a magistrate, if he believes there has been an infringement of copyright. Section 65 of the copyright law has been changed to provide a punishment of up to two years' imprisonment or a suitable fine for an individual who "knowingly makes, or has in his possession, any plates for the purpose of making infringing copies."

Video library owners reacted strongly and swiftly to these modifications. Their organization, All-India Video Clubs Association, contended that the owners' collection of cassettes, rendered illegal by the act, was actually legitimate because it was purchased before the act was promulgated. They emphasized that they had already paid taxes on their cassettes and, if the act was enforced, they would have to dump thousands of valuable tapes (Narwekar, 1985, p. 6).

Among other laws modified in 1984 to account for video were the Cinematograph Act, 1952, and the Cinematograph Certification Rules, 1983. Section 7 of the former provided severe punishments for a person who "exhibits or permits the exhibition of a video film in any place." The penalty included a prison term of three months to three years and a fine of Rs20,000 to Rs1 lakh (U.S.$1,400–$7,000). For regular violators, a further fine can be imposed of up the Rs20,000 for each day the offense continues. The Cinematograph Certification Rules required registration of all video films meant for exhibition, stipulating that "a copy of the censor certificate showing the serial number, the category of the certificate and other relevant details be pasted on every video cassette as well as the outer case."

Because laws affecting video (except for those of censorship) are under the jurisdiction of individual states, pressure has been brought upon their officials to take remedial action. There also have been calls for the centralization of legislation, but this would require a constitutional amendment. Of course, some states are more conscientious than others about this duty. In Karnataka, for example, the Film Chamber of Commerce, in 1983, had enlightened the state government on sections of the States Cinema Act 1964, that disallowed exhibition of cinematographs, except in licensed places (Gopal, 1986, p. 124). The next year, the Karnataka government made applicable to video parlors the same rules of cinema houses. In Tamil Nadu, where the head of state is a movie idol and, therefore, protector of the film industry, video parlors are to be licensed in the same manner as liquor stores; this decision resulted in a spate of legal cases in which video parlors asked for more time to acquire a license. Even some government-owned undertakings in Tamil Nadu, such as long-distance buses and airports, have not abided by the law, as they show videos without a license. In 1984 Maharashtra state proclaimed that special courts would be set up to solve video piracy problems.

Whether central- or state-government enforced, the effectiveness of these Indian regulations remains to be seen, especially in rural areas, where video merchants have been ingenious in their attempts to escape the authorities. How ingenious? In one South Indian state, a mosque in a Muslim burial ground was used as a video parlor because of its isolation and immunity from the police. In other cases, parlors are mobile for the same reason (Rao, 1986, p. 6).

THE IMPACT OF VIDEO

What becomes obvious in assessing the impact of home video is that far more concern is voiced about economic than sociocultural implications. Among the sectors affected by video are film industries, national television services, legitimate video producers and distributors, government taxation bureaus, and sociocultural institutions such as religion, education, family, and work.

Film Industries

Although not commonly known, Asia has had some of the most profitable and prolific film industries in the world. At the top is India, which has led the world in the production of feature films for years, while countries such as the Philippines, China, Taiwan, and Hong Kong also have had thriving film businesses. Virtually every Asian nation is a producer of feature-length films.

Since the advent of the 1980s, Asian cinema has suffered losses in box office attendance and revenues, as well as export markets, culminating in a reduction in the number of films produced and the closing of hundreds of cinema houses. Indian and Philippine film personnel claim their box office attendances are down 50 percent (interview, Lamberito Avellana, Quezon city, Philippines, August 16, 1986), one producer venturing that 95 percent of Indian films never reclaim their costs. Other countries, such as Taiwan, Singapore, and Sri Lanka estimate 20 to 30 percent drops. By 1981, Singapore had already lost 30 to 40 percent in box office attendance (the figure is probably higher now, since 75 percent of Singapore households have VCRs); equally affected is the distribution of 16 mm films rented out to homes, clubs, and societies. Sri Lanka has experienced a 30 percent drop at the box office since 1983; but television, more than video, and higher theater prices are blamed (Kurukulasuriya, 1983, p. 69). Officials of the Indonesian National Film Council have been worried about video replacing celluloid technology (Lubis, 1983, p. 54), while others have claimed video has helped the local films. The latter rationalize that because video piracy has focused on imported movies, box office drop-offs have affected United States and European imports, leaving movie houses to the local film producers ("Indonesian vidpiracy aids local picbiz," 1983, p. 12).

The number of films produced and theater outlets available has diminished in most countries. In Bangladesh, fifty full-length films were produced yearly until 1978; the figure gradually dropped to 36 by 1981, and even lower later. Malaysia, which used to produce about twenty-four films yearly, does eight or fewer now (interview, Dato L. Krishnan, managing director, Gaya Filems, Kuala Lumpur, Malaysia, November 17, 1986). Exporters of films such as India and Hong Kong have also felt the pinch of video.

As is often said, the public's preference for video arises because of video's capacity to offer more varied and explicit fare for less money. In Taiwan, for one-twentieth the cost of film attendance, an entire family can view a videocassette for an evening; in Malaysia, a family can view a video for an entire week with the money required for one person's film ticket for an evening. A Filipino film producer said twenty people can rent cassettes for less money than the price of one theater ticket (interview, Fyke Cinco, director, in Quezon City, Philippines, August 16, 1986).

As a result, film producers and distributors paint a bleak future for their operations. For example, in Taiwan, young producers such as Jen Wan do not feel they have a chance. According to him:

> Video is killing the film industry. We new directors release our movie today and two days from now, it is on video all over. The legal video of our film is released two weeks to a month after the original release. The producer only gets NT$200,000 (U.S.$6,000) for copyright use of the films on video (interview, Jen Wan, in Taipei, August 7, 1986).

Others point out similar frustrating situations where illegal video-cassette versions of movies were on the market before the film hit the theaters (interviews, Dan Tchii and Richard Tu, managing director, China Educational Recreation, in Taipei, August 4, 1986). The director-general of the Government Information Office, as well as others in Taiwan, said that the film industry must come up with incentives to entice an older audience back to the theaters. Among these would be more modern and cleaner theaters that would offer a variety of films simultaneously (interview, Chang King-Yuh, Taipei, August 6, 1986).

Concerns of the Hong Kong film industry are with its overseas, rather than home, markets. Top officials at two of the largest film producers, Cinema City and Golden Harvest, verified, for reasons given earlier, that Hong Kong residents are not voracious viewers of video, so drop-offs in local box office attendance are not the anticipated problem. The impact on their foreign markets—especially in countries with large Chinese-speaking populations and the Chinatowns of the United States and elsewhere—has been devastating. As a partial remedy, both production houses now release their films concurrently in Taiwan and elsewhere to counter piracy. Golden Harvest spends hundreds of thousands of United States dollars to counteract pirates, using private investigators to locate copyright infringements, after which suits are brought against the vendors (interview, David Chan and Peter Lam, Hong Kong, August 12, 13, 1986).

In Southeast Asia, Filipino veteran director Lamberto Avellana predicted the death of the local film industry without adequate supervision of video (interview, Quezon City, Philippines, August 16, 1986). In recent years, the movies in the Philippines have lost 40 percent of their collective grosses, while video operators gross yearly earnings of P600 million (U.S.$30 million). Until 1978, a million Filipinos watched a movie daily; since then, attendance has dwindled to the extent that 300 theaters have closed (Velarde, 1986, p. 15). Both foreign and Tagalog movies have been affected, 10 to 20 percent of the pirated films being the latter (Giron, 1986, p. 443). Foreign sales of Tagalog films have nearly ceased, whereas at one time, 50 to 100 were sent abroad yearly.

Philippine distributors are angered when they pay U.S.$10,000 to $250,000 (the average is $20,000) for rights to a foreign hit, only to find that pirates have already scooped them. Sixto Dy, president of the Integrated Motion Picture Importers and Distributors Association, said similarly about local films:

> A producer spends at least P1.5 million [U.S.$75,000] to put a movie—the cheapest kind—together. Bigger studios probably spend P5 million [U.S.$250,000] per project, including promotions. Imagine their horror when they take the movie to the provinces, and the people there say they have seen it on video (Velarde, 1986, p. 15).

Other movie personnel mentioned the provinces as the place where pirates dominate. Film director Fyke Cinco said that the pirates

> knock us off in the provinces, because we release films slowly—one print going to four or five places. Maybe we take a month to get to Cebu, even longer to Davao. So the pirates have a chance to copy the films. The idea of slow release has been to maximize the audience in one region before marketing in the next (interview, Fyke Cinco, Quezon City, Philippines, August 16, 1986).

Another director, Peque Gallaga, said that because rural audiences have more of a penchant for pornography, the pirates—and even some producers—splice films to insert explicitly sexual scenes (interview, Peque Gallaga, director, Quezon City, Philippines, August 20, 1986).

Video operators in Manila, as mentioned earlier, are well organized and vociferous. If they think they have been aggrieved, they write President Corazon Aquino, pointing out that movie personnel are, and always were, supportive of deposed president Marcos. In their talks with movie people, they have insisted on lowering the waiting time between a film's release and its video duplication. Movies normally take six months to recoup their investment, yet video distributors demand that they be allowed to duplicate them within three, not six months (Velarde, 1986, p. 15).

Among the many ways that films are illegally copied in the Philippines are by means of prints released by dishonest Board of Censors personnel; at special previews; through technicians who "lend" copies to duplicators for three hours for fees of P5,000–P10,000 (U.S.$250–$500); and via runners who shuttle reels to and from theaters that show the same film on alternating schedules.

The Malay film business, which was fairly important in the 1950s with works of P. Ramlee and others, is definitely at death's door.

Television and home video, as well as changing leisure patterns, have taken their toll, and the 400 movie theaters of the 1950s have shrunk to fewer than 100 (interview, Dato L. Krishnan, Kuala Lumpur, Malaysia, November 17, 1986). Theater attendance continues to plummet. A 1978 poll, for example, showed that of those surveyed, 19 percent had been to a theater within the week; six years later, the figure was 8 percent (Tan, 1987, p. 37). Hindi films, as well as Chinese-language films made in Hong Kong and recorded from television screens there, are especially popular on home video (Koh, 1982, p. 47).

All South Asian film producers balk at the real and potential damage of home video. Smaller industries on the subcontinent have been nearly killed by a prevalent mentality expressed by one Pakistani: "Why go to the movies when fifteen days after *Time* magazine reviews a United States movie, you can rent it here?" (interview, Abbas Muzaffar, Lahore, Pakistan, July 12, 1984). In Nepal and Pakistan, theaters have been shut or converted into shopping centers because of home video; and in Bangladesh, foreign videocassettes have been detrimental in pointing up the shortcomings of the local films. Some critics have questioned whether the country needs a film industry, believing that the banned Indian and Pakistani films, which are of better quality, should be imported freely, instead of wasting time and money on cheap local versions (Kamaluddin, 1982, p. 98).

Sri Lanka has experienced dramatic losses in cinema attendance since 1980 (as many as 20 million fares for Sinhala and 10 million for Tamil films), mainly because of the advent of television, but partly because of home video. Recently, the film industry lobbied to stop the development of video theaters. Thus far, the blockage has been successful, as the video entrepreneurs hesitate to take the financial risk as long as film people oppose it. The head of Sri Lanka television, siding with the video exhibitors, said:

> I told the film people that technology is something you have to live with and that video has its own advantages. With video, history is repeating itself. The old Film Commission Report of the 1960s was meant to help the film industry, but instead, brought in a monopoly situation. If restrictive policies are applied to video, a similar situation will result. For one thing, no country has stopped piracy. The reality of it is that there will be police taking bribes and other types of corruption. It is far better to legalize video so that the film people can make a living from it (interview, Anura Goonsekera, Singapore, November 20, 1986).

Video is particularly worrisome in India, which has an annual production of nearly 800 films and a turnover of U.S.$800 million

(Suraiya, 1983, p. 80). Already by 1983, the U.S.$700 million investment in Indian film was matched by an equal investment in video (Sarathy, 1983, p. 53) further increasing producers' agitation and paranoia. Video piracy has been so rampant and fast that, as mentioned before, a movie was on thousands of cassettes the day after it was released in London and a day before its showing in India (*India Today*, September 15, 1984, p. 131). One producer was said to be so paranoid about the possibility of piracy that he tucks the master print of his latest film under his mattress at night.

Because of the corruptible nature of the Indian film industry, not much sympathy has been shown producers and directors. For years, a major conduit for the flow of undeclared "black money," the industry is not noted for its business ethics, often being financed by smugglers and racketeers. Also, as one source said, charges of video piracy "have an ironic echo of poetic justice, coming as they do from an industry notorious for its flagrant plagiarism" (Suraiya, 1983, p. 80).

One writer showed that, at home, Indian films prosper in spite of the enormous influx of video. During the first six months of 1985, all A-grade films (i.e., big budget) opened to 90 percent or more earnings; and even regional ones released in Bombay did 90 percent business during opening week. He explained that the movies were not hurt as much as was feared because the lower classes, the main patrons of Indian cinema, could not afford home video. He added that the video circuit only mops up the excess demand for Hindi films that cannot be satisfied by the 11,000 theaters (Narwekar, 1985, p. 6).

But, the real loss is the collapse of the entire foreign market for Indian film, both in revenues and numbers of films exported. Whereas in 1979–1980, India exported 199 films, in 1982–1983, the market accommodated only 51, and the following year, about half that number. Countries such as Kenya, which at one time imported many Indian films, quit doing so altogether. Because video looms large throughout the world, Indian exports to other regions have been affected. By 1985, the drop in value of exported cinema to the United States and Canada was 96.7 percent; Malaysia, 93.5 percent; Great Britain and Ireland, 89.2 percent; the West Indies, 88 percent; the Middle East, 76.4 percent; and Sri Lanka, 65.4 percent (Narwekar, 1985, p. 6).

Television and Legitimate Video

Censorship policies affecting national television services, and program development and use for government propaganda, have been partially to blame for diverting the public interest to home video. In country after country, audiences have been dissatisfied with regular television programs. Muslim group representatives in Indonesia have chastised TVRI for being "too government oriented," for using nothing that conflicts with the government ideology (*Straits Times*, September 4, 1984, p. 8); similar complaints of dull, propagandistic, or irrelevant programming have been heard regularly throughout Asia (interview, Lan Tsu-Wei, reporter, *United Daily News*, Taipei, August 7, 1986).

Television ratings have shown that, with the alternative of videocassettes, on-the-air programs have suffered. Before video was pervasive in Taiwan, the top-rated shows drew 60 to 70 percent of the potential audience; they fell to 40 to 50 percent by the mid-1980s. The head of Taiwanese television said that other leisure activities also affected television viewing, but conceded that video was the biggest culprit. She said that television stations in Taiwan have attempted to win back audiences by offering programming that video is not known for—better children's shows and Taiwanese soap operas. In addition they are attracting first-run films for television (interview, Sunshine Kuang, Taipei, August 8, 1986).

In Malaysia, Singapore, and Thailand, a new television service was introduced to win back audiences. From 1982 onwards, Malaysian government officials have spoken out strongly against the effects of video upon television. They have complained that the public no longer watches and, therefore, government developmental programs have not been seen; that national integration has been affected as the Chinese have watched imported videocassettes in their languages, rather than Television Malaysia shows in the national language; and that content contrary to Muslim values has been seen on video. The Chinese viewing problem probably has been uppermost in their minds, as they have worried about reopening Chinese exposure to Chinese culture through video. The authorities found themselves in the dilemma of either making Television Malaysia attractive to the Chinese and compromising their national integration objectives in the process, or permitting the Chinese to watch foreign-produced video, again at the sacrifice of national integration goals (Kob, 1982, p. 46). Initially, plans were made to improve the content of Television Malaysia; but then, in 1984, a third channel, TV-Tiga, was opened, with the main purpose of competing with video. Started by private interests tied to the United Malays National Organisation, the dominant party in the ruling coalition, TV-Tiga obviously had governmental interests at heart. From the beginning, the station announced its purpose as drawing back viewers, claiming "good TV drives out bad video."

Thai television has felt the impact of video. The amount of revenue lost is not known, but the TV audience in 1986 dropped by 10 to 15 percent. Channel 9 reacted in 1987 by implementing cable television (Hamid, 1987, p. 33).

Legitimate video production and distribution businesses have trouble competing with pirates, who can offer lower prices because they do not pay taxes, and have very little overhead and no advertising or promotional expenditures. In the Philippines, where 99 percent of video is estimated to be in the hands of illicit traders, one of the few legitimate concerns—Roadshow Video—felt "deeply hurt in competition with the network of well-organized pirates." Another concern, Viva Home Entertainment, planned to compete by establishing its own chain of video stores and offering more local productions. In late 1986, Viva had 400 film titles, 75 percent of which were locally done.

Indonesia's P. T. Trio Video Tara, a large videocassette firm created in 1979, in one three-month period brought out 108 titles, mostly foreign films that it purchased the rights to and duplicated ("Cassette scandal," 1980, p. 21). By 1986, other legitimate companies such as Super Picture Video entered the market, but all predicted disastrous consequences because of video piracy.

Potential Government Revenue

Asian governments have lost millions of dollars in tax revenues because of illegal videocassettes and VCRs. This is particularly the case in the Philippines, where cinema is the most heavily taxed industry. Already taxed 42 percent on earnings, Philippine movie producers and distributers pay additional taxes such as amusement, cultural, municipal, film stock, and flood (levied to improve drainage in the country). One film writer explained that films are singled out for heavy taxation because they are the people's medium and, as such, have the potential to bring in large amounts of money (interview José F. Lacaba, Quezon City, Philippines, August 22, 1986). To give an indication of the losses, the government said that as early as 1979, it lost U.S.$36 million in revenue from the sale of smuggled VCRs and cassettes. In 1981, it stood to lose U.S.$2.1 million in exhibition taxes from Metro Manila theaters, because of a 30 percent reduction in gross billings from foreign films. The previous year, taxes of U.S.$2.9 million on foreign film sales, U.S.$3 million on Filipino movies, and over U.S.$1.3 million in film industry income taxes had been collected.

In Bangladesh, where the film industry generates U.S.$15.6 million in entertainment taxes, the prediction was that 30 to 40 percent of that would be lost to piracy. In 1982, film officials recommended to the government that VCR showings be regulated, cassettes be imported, and entertainment taxes levied to give the government needed revenue and raise the prices for VCR operators, thus curbing their competitive edge (Kamaluddin, 1982).

Pirates deprive the Taiwan government of the import duty of U.S.$1,000 to $2,000 on foreign films, as well as the box office tax of 35 percent and the VCR sales duty of 15 percent (interviews, Danny Tchii and Sunshine Kuang, Taipei, August 5, 8, 1986). India's central and state governments passed adhoc legislation in the 1980s to contain video, especially after realizing that in early 1984, New Delhi's entertainment tax income had dropped by U.S.$154,000 over the previous month. In 1982, the northern state of Uttar Pradesh lost U.S.$6 million in tax revenue because of the prevalence of illegal video.

Sociocultural Factors

The impact of video upon individual and societal roles, expectations, and values in Asia, has not come under very close scrutiny. Unlike those of an economic nature, sociocultural influences are not easily identifiable and are even less provable. For example, when asked about the effects of video, most Indian respondents to a survey conducted in Delhi and a neighboring settlement "could not say," although 18.3 percent said it was "bad for children"; 16.4 percent, a luxury at home; and 15.6 percent, "wastage of time" (Yadava, 1986, p. 3). With a paucity of hard data, much of what is said about the sociocultural impact of videocassettes is impressionistic and even anecdotal, but nevertheless worth listening to.

A cross-section of the types of possible effects is represented in a statement made by the head of television in Taiwan, who said:

> This office is concerned with the cultural impact of video. Some people say we should treat video like books and let the public make its own decision on what they see. But, it is different with books as video has sound and visuals. Video has a more severe impact than books, but we will leave it to scholars to find out what that is. Some people say maybe it is better not to ban foreign materials, but to give more variety of Chinese products showing our culture. But people choose the foreign video. Newspapers and magazines now question video's cultural impact. Maybe video is a good stimulus for the imagination? But, from our point of view, it is kind of dangerous, as people watch too much fantasy and do not touch on reality. People can get the wrong idea of our life-style. Fast-paced video makes young people more anxious and impatient. Life is expectable but video too often makes one have different expectations. In the United States, a kiss in public is okay, but here it is strange. Living together without marriage and homosexuality are against our culture, but if

viewers see it too often on video, they think it is normal (interview, Sunshine Kuang, Taipei, August 8, 1986).

Among the social institutions particularly sensitive to the effects of video are religious organizations, especially those in Islamic nations. For example, the drive towards a Muslim culture by Malaysia, Indonesia, or Pakistan, has faced setbacks as the control of morality has been taken from the broadcaster and put into the home. Besides enticing Muslims with violent and sexually explicit material, home video has also kept them occupied (e.g., 38 percent of Malay video viewers according to one survey) while religious programs are on television (Ganley & Ganley, 1986, p. 93). The Roman Catholic Church, especially in the Philippines, has reacted negatively, at times, to pornographic video. Video in Malaysia has become popular enough to warrant a place among worship paraphernalia. Replicas of video equipment and cassettes often are among the offerings given with prayers to the dead.

Indonesian parents, as well as those in other countries, worry about the connections between home video and their children. They complain that children learn about sex too early, are not educated well because they substitute video for homework, and are saturated with violence and other dubious moral values through the medium (Lubis, 1983, p. 54). Thai psychiatrists believe that children in that country have become more physically aggressive from viewing television and video.

In other countries, authorities are disturbed that literacy, reading habits, and educational levels have suffered with video's thrust. In China, where most television programs and films are educational, the government is worried about the public's exposure to fictional material that it has not cleared, and that when viewed by the uneducated, may be taken for reality. In Singapore, where government ministers often stress the value of reading, the deleterious impact of video upon this pastime has been discussed (Koh, 1982, p. 48). For years, rental libraries (where the public pays a fee to read books on the premises or to take home) were very popular in Taiwan. With the dominance of television and video, many libraries have dried up; most of the 1,000 remaining in Taipei are in their owners' homes, run by retirees and housewives on a part-time basis (Huang, 1986, p. 3). The population's preference for video in Indonesia has affected what is being translated and published in book form (Suryadinata, 1985, pp. 38–39).

Video definitely has had an impact upon how some Asians socialize. In India, video parties are faddish among the elites, as are videocassettes of weddings, available professionally in edited versions for U.S.$100. Parties often are determined by "who has what to show," and invitation cards among the elites carry the teaser, "drinks and video" (Suraiya, 1983, p. 81). Einsiedel (1986) found that in Philippine villages, video viewing is an important social event. In many cultures, the VCR is a status symbol. Thus, among the upper middle class in India, it is essential on the list of dowry demands; and in many opulent homes, it is garlanded like a household deity and given the place of honor.

Opinions differ on video's role in uniting social classes. Indian efforts over the years to declass media by removing them from small elite groups and making them available to the masses, have faced setbacks from video. Agrawal (1986, p. 39), for example, said video viewing becomes kin- or class-oriented, with groups that have "a class character similar to those observed in nineteenth-century rural India. These groups had common social and economic backgrounds." As a result, he predicted video will create a set of "small, homogeneous, socially visible elite groups" that will lead to "forced privatization of recreation" and media inequality. Agrawal believed that the information gap will widen because television and video viewers will have access to different types of information; furthermore, the Indian rich will be drawn into the orbit of the international rich and be more removed from the Indian poor. Therefore, according to Agrawal, video will unite Indian elites and strengthen solidarity with elites throughout the world. At the opposite extreme, *Asiaweek* (May 4, 1984) called video the great social leveler in India, saying that,

> a roadside barber from Bombay's impoverished suburb . . . now has the same viewing privileges as the industrialist in posh Malabar Hill who can afford his own VCR.

Perhaps video has very slightly closed the information gap between men and women in some areas. For example, as South Indian males have gone to the labor market in the Middle East, and their watchful eyes are not ever present, opportunities have opened up for wives and daughters to watch video at home and in parlors. Rao said that in some Muslim parlors, separate rooms or special films are provided to women. She added:

> In one particular parlor, the owner has two rooms in his house for viewing. He shows the same film in the two rooms—one for women, one for men. Before, Muslim men and women would never go to a social function together. Now they do. The women go to the same show but separately. Most parlors have separate times for women— in the afternoon (interview, Leela Rao, Singapore, November 23, 1986).

In these parlors, Muslims and Christians also watch together, something they normally would not do in theaters.

Leisure and work habits are affected by the ready availability of home video in some areas. In Taiwan, a popular magazine reported low working morale in the mornings, attributable to late-night VCR. Studies showed that 40 to 60 percent of video viewers watch after television goes off the air at night. One top-level corporate manager described the phenomenon as the "VCR syndrome" (Wang, 1986, p. 374). Reports have also been made of Indonesian fishermen being reluctant to go to sea in the evening because they would miss television and video viewing. In China, video has created a new line of work, according to one source (Ganley & Ganley, 1986, p. 87). The government, to keep increasingly rich peasants amused, has encouraged peasants to develop "cultural households" that show video, cinema, and theater and accommodate drama troupes, libraries, photo studies, and fine arts services. In five provinces, about 130,000 peasants have chosen the cultural business over farming.

In multicultural, multilingual societies, the evidence on the impact of video again is far from conclusive. One suggestion was that video would be extremely useful in a country such as India, with its multiplicity of languages and cultures, to capture localness. In Indonesia, however, officials worried that the heavy viewing of Kungfu movies on VCRs by local Chinese would reorient them to China and impede Indonesia's national culture. Similarly, the high video viewership among Malaysian Chinese has been considered anathema to assimilation of that ethnic group.

Other critics have noted that video viewing can raise consumer expectations beyond their realization levels. Modes of living, including those revolving around fashion, food, automobiles, houses, and so on, are reflected in foreign- and urban-created video. In some cases, where television programs have been illegally taped and sold, commercials for high-priced products have not been deleted. Such tapes can be very troublesome in a country such as Indonesia, where commercials have been banned from national television for years.

Of course, the problem most often heard worldwide is the cultural invasion that foreign-produced media (videocassettes among them) promote, leading to the decimation of local traditional and folk forms, homogenization, and lack of cultural autonomy. A 1983 Reuters survey of fifteen Asian countries indicated that only Japan did not have VCR-inspired cultural or other worries (Ganley & Ganley, 1986, p. 99). In Thailand, the media have carried on a dialogue about the long-term effects of heavy exposure to foreign culture, pointing out that television has been used as a carefully controlled unifier and purveyor of

Thai culture, while the VCR has countered this influence (Ganley & Ganley, 1986, p. 99). To phase out the Japanese impact from colonial days, and to put more emphasis on local content, Taiwanese authorities banned Japanese drama (except cartoons) from television. But the VCR has eroded this independence.

Discussing VCRs and the cultural invasion, Wang (1986, p. 377) concluded that the situation in Taiwan is peculiar:

> . . .cultural invasion is feared, but because of piracy, some "cultural imperialists" seemed to have profited little from the growth of VCR in Taiwan;
>
> — on paper, the government has control over all forms of video products, but thanks to VCR, audiences are free to watch everything provided by rental shops;
>
> — VCR has helped to satisfy the needs of audiences, but in the long run, what consequence it will bring to the local is not clear;
>
> — the government has targeted for further VCR growth, but the more VCRs the more widespread the problems will be should the present situation continue.

Finally, political uses of videocassettes are feared in Asian countries because of their potential effects upon society. Thus India keeps an eye on the video that is circulated there, as it does not want to upset neighboring societies; other countries worry about communist infiltration through video. During the 1984 elections in Malaysia, an important debate on fundamentalist Islam was called off by the king, to the relief of neighboring countries (especially Indonesia), who fretted that a videotaped TV debate such as this could hurt their own domestic situations.

NOTES

1. All dollar conversions are as of mid-1987.
2. One author claimed four types of sources for Indian video: (1) original prints made by the few licensed video recording companies; (2) prints from abroad recorded from the original film print in Hong Kong, London, or the Gulf; (3) first-copy prints made from the original print in India; and (4) camera prints recorded clandestinely during a film's screening in a theater (Gopal, 1986, p. 13).
3. Although Japan is not within the purview of this chapter because it is not part of the Third World, some figures on the prevalence of cassette piracy

there are given to fill out the Asian picture. After a tour of Japan in late 1986, United States movie representative Jack Valenti estimated that there were 6 million to 7 million pirated cassettes, representing 40 to 50 percent of all video revenues and a loss of U.S.$519 million to producers. Valenti also gave statistics on the hugeness and growth of video in Japan: from 1981 to 1982, the country had a 107 percent growth in VCRs; 1982–1983, 106 percent; 1983–1984, 45 percent; and 1984–1985, 14 percent. Videodiscs grew by 10.1 percent in 1983, 43.1 percent in 1984, and 57.1 percent in 1985. Over 6,000 videocassette rental firms exist in Japan (*Variety*, October 8, 1986).

APPENDIXES

4.A: The Enforcement Rules of the Broadcasting and Television Law (Taiwan)

We have discussed various means utilized by governments in the Third World to control the sale and rental of video cassettes. The Broadcasting and Television Law in Taiwan has been broadened to include video tape. What follows is the attempt of Taiwan's Government Information Office to provide rules for the distribution of cassettes. See Appendix 3.A for the much more detailed guidelines offered by Saudi Arabia.

Approved on December 24, 1976, by an Executive Yuan letter, (65) wen 10933, and released by a government Information Office decree, (65) mao po 12954.

Approved on November 2, 1979, by an Executive Yuan Letter, (68) wen 10967, and released by a Government Information Office decree, (68) yu kuang 13957.

Approved on April 18, 1983, by an Executive Yuan letter, (72) wen 6768, and released on April 25, 1983, by a Government Information Office decree, (72) yu kuang 05366.

VIDEO-TAPE CONTROL

Article 33
The GIO may authorize the provincial (municipal) and county (city) government to control and regulate the designated items of video-tape programs.

Article 34
The video-tape programs mentioned in paragraph Ten of article 2 of this Law include video discs and other forms of products that may be produced on a television screen through electronic scanning. But computer programs are not included.

Article 35
In the case of a video-tape program, the title of the program, and the name, address of the producing company or concern, the names of its

responsible personnel, and the number of the registration certificate issued by the GIO shall be inscribed on the cassette and on the tape.

The holder of a video-tape which does not carry an inscription to indicate the above-mentioned items or the contents of the program—or if the indication is inconsistent with the facts—shall be considered as owner of the video-tape.

Article 36
An application together with the video-tape shall be filed with the GIO for examination and issuance of a certificate before distribution or broadcasting if the contents of the video-tape are so designated by the GIO or if the video-film, the certificates of copyright and distribution rights shall also be presented in the application.

Upon examination and approval, the applicant shall indicate the serial number of the certificate on the cassette and the tape, and the GIO may keep a copy of the video-tape for reference.

A taped video program purely for educational purpose shall be regarded as teaching material and shall be sent to the Ministry of Education to be dealt with in accordance with related regulations.

Article 37
The contents of a video-tape program shall not include any contingency relating to Article 21 of this Law. If it is a film, its screening shall not be prohibited. Nor shall its approved items be changed.

Article 38
If a video-tape program contains advertising, it shall be examined according to regulation. If approved, the GIO may keep a copy for reference.

If television advertising approved by the GIO is included in a video tape program, the advertising is exempt from examination in accordance with the provisions of the preceding paragraph so long as the broadcasting certificate of the given advertising is valid, but if the GIO recalls the advertising for reexamination and decides to prohibit its broadcast, the advertising shall be deleted from the tape.

Article 39
If only one audio-visual machine is used for broadcasting a video-tape program, the broadcast shall, in principle, be made at home. If it is to be broadcast publicly, the location of broadcasting must be a site set up according to law by a film screening establishment, and the video tape

being screened must have a certificate obtained in accordance with the provisions of Article 36. This does not apply in connection with any of the following conditions:

1. An educational video-tape may be broadcast in an educational location.
2. A professional video-tape may be broadcast in a professional location.
3. If the broadcast is a preview for customers, free of charge, or if it is for testing the broadcasting equipment, the broadcast shall be made at a business location.

If the foregoing provisions are violated, the broadcaster shall be punished as one who has installed his broadcasting system in violation of lawful procedures.

Article 40
The import or export of video-tape programs has to be approved. But if the contents violate Article 37, the tape program will be confiscated.

4.B: Presidential Decree 1987: An Act Creating the Videogram Regulatory Board (Philippines)

The Videogram Regulatory Board of the Philippines was created under former President Ferdinand Marcos. The following Presidential Decree No. 1987 not only provides the structure for the VRB, but also gives some idea of the type of control the president had in mind. Because Marcos was determined to control the flow of printed and broadcast information to citizens, he saw the VCR as a threat to political stability, and thus to his government.

Whereas, the proliferation and unregulated circulation of videograms including among others, videotapes, discs, cassettes of any technical improvement or variation thereof, have caused a sharp decline in theatrical attendance by at least forty percent (40%) and a tremendous drop in the collection of sales, contractor's specific amusement, and other taxes, thereby resulting in substantial losses estimated at P450 Million annually in government revenues;

Whereas, videogram establishments collectively earn around P600 Million per annum from rentals, sales and disposition of videograms, and such earnings have not been subjected to tax, thereby depriving the Government of approximately P180 Million in taxes each year;

Whereas, the unregulated activities of videogram establishments have also affected the viability of the movie industry, particularly the more 1,200 moviehouses and theaters throughout the country, and occasioned industry-wide displacement and unemployment due to the shutdown of numerous moviehouses and theaters;

Whereas, in order to ensure national economic recovery, it is imperative for the Government to create an environment conducive to the growth and development of all business industries, including the movie industry which has an accumulated investment of about P3 Billion;

Whereas, proper taxation of the activities of videogram establishments will not only alleviate the dire financial condition of the movie industry upon which more than 75,000 families and 500,000 workers depend for their livelihood, but also provide an additional source of revenue for the Government, and at the same time rationalize the heretofore uncontrolled distribution of videogram;

Whereas, the rampant and unregulated showing of obscene videogram features constitutes a clear and present danger to the moral and spiritual well-being of the youth, and impairs the mandate of the Constitution for the State to support the rearing of the youth for civic efficiency and the development of moral character and promote their physical, intellectual, and social well-being;

Whereas, civic-minded citizens and groups have called for remedial measures to curb these blatant malpractices which have flaunted our censorship and copyright laws;

Whereas, in the face of these grave emergencies corroding the moral values of the people and betraying the national economic recovery program, bold emergency measures must be adopted with dispatch;

Now, THEREFORE, I FERDINAND E. MARCOS, President of the Philippines, by virtue of the powers vested in me by the Constitution, do hereby decree:

SECTION 1. *Creation.* There is hereby created an office to be known as the Videogram Regulatory Board, hereinafter referred to as the BOARD, which shall have its principal office in Metro Manila and shall be under the Office of the President of the Philippines. The BOARD shall regulate the importation, exportation, production, reproduction, distribution, exhibition, showing, sale, lease or disposition of videograms including, among others, videotapes, discs, cassettes or any technical improvement or variation thereof in accordance with such rules and regulations to be adopted by the BOARD.

SECTION 2. *Composition of the Board*. The BOARD shall be composed of a Chairman, a Vice-Chairman, and eighteen (18) members who shall all be appointed by the President of the Philippines to serve for a term of one (1) year, unless sooner removed by the President for any cause; *Provided*, That they shall be eligible for reappointment after the expiration of their term. If the Chairman or Vice-Chairman or any of the members fails to complete his term, any person appointed to fill the vacancy shall serve only for the unexpired portion of the term of the Board member whom he succeeds.

No person shall be appointed to the BOARD, unless he is a natural-born citizen of the Philippines, not less than twenty-one years of age, and good moral character and standing in the community; *Provided*, further, That at least five (5) members must be members of the Philippine Bar.

The Chairman, Vice-Chairman, and members of the BOARD shall be entitled to transportation, representation, and other allowances which shall in no case exceed Five Thousand Pesos (P5,000.00) per month.

SECTION 3. *Powers and Functions*. The BOARD shall have the following powers and functions:

1. To supervise, regulate, grant, deny, or cancel permits for the importation, exportation, production, copying, sale, lease, exhibition, or showing of videograms including, among others, videotapes, discs, cassettes, or any technical improvement or variation thereof;

2. To approve or disapprove, delete objectionable portions from and/ or prohibit the importation, exportation, production, copying, distribution, sale, lease, exhibition, or showing of videograms, including, among others, videotapes, discs, cassettes, or any technical improvement or variation thereof, which, in the judgement of the BOARD applying contemporary Filipino cultural values as basic standard, are objectionable for being immoral, indecent, libelous, contrary to law or good customs, or injurious to the prestige of the Republic of the Philippines or its people, or with a dangerous tendency to encourage or fan hatred, the commission of violence or of a wrong or crime, such as but not limited to:

    i. Those which tend to incite subversion, insurrection, rebellion, or sedition against the State or otherwise threaten the economic and/or political stability of the State;

    ii. Those which tend to undermine the faith and confidence of the people in their government and/or duly constituted authorities;

    iii. Those which glorify criminals or condone crimes;

iv. Those which are libelous or defamatory to the good name and reputation of any person, whether living or dead;

v. Those which serve no other purpose but to satisfy the market for excessive violence or hard-core pornography;

vi. Those which tend to abet the traffic in and use of prohibited drugs; or

vii. Those which commit direct or indirect contempt of any court of justice or quasi-judicial tribunal whether any litigation on the subject of the video is pending or not before such court or tribunal.

3. To classify all videograms into categories such as "For General Patronage," "For Adults Only," or such other categories as the BOARD may determine for the public interest;

4. To close video theaters and other similar establishments engaged in the public exhibition or showing of videograms which violate the provisions of this Decree and the rules and regulations promulgated by the BOARD pursuant thereto;

5. To levy, assess, and collect, and periodically adjust and revise the rates of fees and charges for the work of review and examination and for the issuance of licenses and permits which the BOARD is authorized to grant in the exercise of its powers and functions and in the performance of its duties and responsibilities;

6. To review and examine all videograms including among others, videotapes, discs, cassettes, or any technical improvement or variation thereof, as well as publicity materials or advertisements related thereto, with the end view of making appropriate classification;

7. To deputize representatives from the government and from the various associations in the videogram industry, whose main duties shall be to help and ensure compliance with all laws, rules, and regulations relative to this Decree. For the purpose, the BOARD may constitute such Council or Councils composed of representatives from the government and the videogram industry as may be appropriate to implement the objectives of this Decree. The BOARD may also call upon any law enforcement agency for assistance in the implementation and enforcement of its decisions, orders and rules and regulations;

8. To cause the prosecution, on behalf of the People of the Philippines, of violators of this Decree and the rules and regulations promulgated or issued by the BOARD;

9. To promulgate such rules and regulations as are necessary or proper for the implementation of this Decree and the accomplishment of its purposes and objectives. Such rules and regulations

shall take effect after fifteen (15) days following their publication in newspapers of general circulation in the Philippines;

10. To prescribe the internal and operational systems and procedures for the exercise of its powers and functions including the creation and vesting of authority upon sub-committees of the BOARD for the work of review, examination, or classification and other related matters; and

11. To exercise such other powers and functions as may be necessary or incidental to the attainment of the purposes and objectives of this Decree, and to perform such other related duties and responsibilities as may be directed by the President of the Philippines;

SECTION 4. *Executive Officer.* The Chairman of the BOARD shall be the Chief Executive Officer. He shall exercise the following duties and functions:

a. Execute, implement, and enforce the decisions, orders, rules, and regulations promulgated or issued by the BOARD;

b. Direct and supervise the operations and the internal affairs of the BOARD;

c. Establish the internal organization and administrative procedures of the BOARD, and recommend to the BOARD the appointment of the necessary administrative and subordinate personnel; and

d. Exercise such other powers and functions and perform such duties as are not specifically lodged in the BOARD.

The Chief Executive Officer shall be assisted by an Executive Director who shall be appointed by the President of the Philippines. The Executive Director shall hold office for a term of one (1) year, unless sooner removed by the President of the Philippines, for any cause.

Unless otherwise provided by law, the Chief Executive Officer shall receive an annual salary of Seventy-Two Thousand Pesos (P72,000.00) and the Executive Director shall recieve an annual salary of Sixty Thousand Pesos (P60,000.00).

SECTION 6. *Registration.* No person, whether natural or juridical, may engage in the importation, exporation, production, reproduction, exhibition, showing, sale, lease, or disposition of videograms unless such person is first registered with and permitted by the BOARD to operate as such. The registration with and permit issued by the BOARD is a condition precedent for securing a business permit or license from the appropriate authorities.

SECTION 7. *Reproduction of Cinematographic Art.* No person registered and permitted to engage in the videogram industry can copy or reproduce any cinematographic art without the written consent or approval of the producer, importer, or licensee of the cinematographic art to be copied or reproduced, and in no case shall any cinematographic art be allowed to be copied or reproduced within a period of six (6) months after it is first released for theatrical exhibition, unless the producer, importer, or licensee agrees to a shorter period.

SECTION 8. *Sale, Lease, or Disposition of Videograms.* No videogram including, among others, videotapes, discs, cassettes, or any technical improvement or variation thereof, shall be sold, leased, or otherwise disposed of unless first registered with the BOARD with the corresponding registration identification or seal in such form and manner as may be provided for by the BOARD.

SECTION 9. *Penalty.* Any person who violates any or all of the provisions of Sections 3, 6, 7, 8, and 10 of this Decree or the rules and regulations to be promulgated pursuant thereto, either as principal, accomplice, or accessory, shall, upon conviction, suffer a mandatory penalty of three (3) months and one (1) day to one (1) year imprisonment plus a fine of not less than Fifty Thousand Pesos (P50,000.00) but not more than One Hundred Thousand Pesos (P100,000.00). Should the offense be committed by a juridical person, the chairman, the president, secretary, treasurer, or the partner responsible therefor, shall be the persons penalized.

The provisions of Presidential Decree No. 968, as amended (Probation Law), shall not apply in cases of violations of this Decree, including its implementing rules and regulations.

SECTION 10. *Tax on Sale, Lease or Disposition of Videograms.* Notwithstanding any provision of law to the contrary, the province shall collect a tax of thirty percent (30%) of the purchase price or rental rate, as the case may be, for every sale, lease, or disposition of a videogram containing a reproduction of any motion picture or audio-visual program. Fifty percent (50%) of the proceeds of the tax collected shall accrue to the province, and the other fifty percent (50%) shall accrue to the municipality where the tax is collected; Provide, That in Metropolitan Manila, the tax shall be shared equally by the city/municipality and the Metropolitan Manila Commission.

The tax herein imposed shall be due, and payable within the first twenty (20) days of the month next following that for which it is due, by the proprietor, seller, or lessor concerned, and such tax shall be determined on the basis of a true and complete return of the amount of gross receipts derived during the preceding month. If the tax is not paid within the time fixed herein above, the taxpayer shall be subject to such surcharges, interests, and penalties prescribed by the Local Tax

Code. In case of wilful neglect to file the return and pay the tax within the time required, or in case a fraudulent return is filed or a false return is wilfully made, the taxpayer shall be subject to a surcharge of fifty percent (50%) of the correct amount of the tax due in addition to the interest and penalties provided by the Local Tax Code.

Any provision of law to the contrary notwithstanding, a city may also levy and collect, among others, any of the taxes, fees, and other impositions that the province or the municipality may levy and collect.

SECTION 11. *Assistance in the Enforcement Functions of the* BOARD. The BOARD may solicit the direct assistance of other agencies and units of the government, and deputize, for a fixed and limited period, the heads or personnel of such agencies and units to perform enforcement functions for the BOARD. The government agencies and units exercising the enforcement functions for BOARD shall, insofar as such functions are concerned, be subject to the direction and control of the BOARD.

SECTION 12. *Organization Pattern; Personnel.* The BOARD shall determine its organizational structure and its staffing pattern. It shall have the power to suspend or dismiss for cause any employee and/or approve or disapprove the appointment, transfer, or detail of employees. It shall appoint the Secretary of the BOARD who shall be the official custodian of the records of the meetings of the BOARD and who shall perform such other duties and functions as directed by the BOARD.

SECTION 13. *Applicability of Civil Service Law.* The BOARD and its officers and employees shall be subject to the Civil Service Law, rules, and regulations; Provided, That technical personnel shall be selected on the basis of merit and fitness to be determined in accordance with such policies and guidelines as may be approved by the BOARD.

SECTION 14. *Auditor.* The Chairman of the Commission on Audit shall be the ex officio Auditor of the BOARD. For this purpose, he may appoint a representative with necessary personnel to assist said representative in the performance of his duties. The number and salaries of the auditor and said personnel shall be determined by the Chairman of the Commission on Audit, subject to the rules and regulations of the Commission on Audit. Said and all other expenses of maintaining the auditor's office shall be paid by the BOARD.

The Auditor shall, as soon as practicability, but not later than three (3) months after the accounts have been submitted to the audit, send an annual report to the BOARD. The Auditor shall also submit such periodic or special reports as the BOARD may deem necessary or proper.

SECTION 15. *Transitory Provision.* All videogram establishments in the Philippines are hereby given a period of forty-five (45) days after the effectivity of this Decree within which to register with and secure a permit from the BOARD to engage in the videogram business and to register with the BOARD all their inventories of videograms, including videotapes, discs, cassettes, or other technical improvements or variations thereof, before they could be sold, leased, or otherwise disposed of. Thereafter any videogram found in the possession of any person engaged in the videogram business without the required proof of registration by the BOARD, shall be *prima facie* evidence of violation of this decree, whether the possession of such videogram be for private showing and/or for public exhibition.

SECTION 16. *Appropriations.* The sum of Three Million Pesos (P3,000,000.00) out of any available funds from the National Treasury is hereby appropriated and authorized to be released for the organization of the BOARD and its initial operations. Henceforth, funds sufficient to fully carry out the functions and objectives of the BOARD shall be appropriated every fiscal year in the General Appropriations Act.

All fees, revenues, and receipts of the BOARD from any and all sources shall be used to augment the funds to support the expenditures needed by the BOARD in the pursuit of its purposes and objectives and the exercise of its powers and functions, and for such other purposes as may hereafter be directed by the President of the Philippines.

SECTION 17. *Annual Reports.* The BOARD shall, within three months after the end of every fiscal year, submit its annual report to the President. The annual report shall include, among others, a statement of the BOARD's accomplishments together with its plans and recommendations to improve and develop its operations and the supervision and regulation of the videogram industry.

SECTION 18. *Separability Clause.* In case any provision of this Decree shall be held or declared invalid or unconstitutional, the validity of the other provisions shall not be affected thereby.

SECTION 19. *Repealing Clause.* Any provision of law, decree, executive order, letter of instructions, or implementation, or other rules and regulations inconsistent with the provisions of this Decree is hereby repealed, amended, or modified accordingly.

SECTION 20. *Effectivity.* This Decree shall take effect after fifteen (15) days following its publication in the Official Gazette.

Done in the City of Manila, this 5th day of October in the year of Our Lord, nineteen hundred and eighty-five.

(Sgd.) FERDINAND E. MARCOS
President of the Philippines

By the President:

(Sgd.) JUAN C. TUVERA
Presidential Executive Assistant

4.C: Video Business: The Birth Pangs of Legitimization (Philippines)

The Philippines has attempted to promote cooperation between producers and video rental stores by creating the Videogram Regulatory Board (VRB). (See Appendix 4.B). The philosophy of this body is to reduce the pirating of material by promoting cooperation between owners of films and video productions and retailers. The following interview was conducted by journalist Eduardo Pacheco and involved Ed Sazon, a VRB member. It appeared in the August 21–25, 1986, edition of *Veritas*, a Philippine weekly magazine.

Over the past weeks, Betamax addicts have complained about the dearth of new titles at their favorite videotape outlets. With virtually no new titles available for home viewing, news that video rentals would increase by as much as 50 percent compounded their woes. In most everyone's mind, the culprit is none other than the newly created Videogram Regulatory Board (VRB). To give our readers a clearer picture of what's happening in the video industry vis-à-vis the movie industry, *Veritas'* Eduardo B. Pacheco talked to VRB officer-in-charge Ed Sazon and some "harassed" video retailers to get their side of the story.

Q. The videotape outlets have reacted to the new Videogram Regulatory Board in unprecedented fashion. Why? Are the rules you are implementing now really that novel?

A. Yes, the VRB formulated new implementing rules and guidelines. The original implementing rules and guidelines were approved by the old board, but they were not publicized. The old implementing rules and guidelines which were approved presumed an amendment of the Presidential Decree so we couldn't enforce them. The new guidelines were published May 12 in a newspaper of general circulation. It became effective May 25, 1986.

Q. How different are the new implementing rules and guidelines?

A. In the old guidelines, we permitted videogram establishments to duplicate tapes without permission from the owners, provided the establishments put up a cash bond. Under the new regulations, we don't permit that anymore. As a person engaged in motion pictures, I would not want government to intervene with proprietary rights. And in terms of cash bonds, because video rights, theatrical rights, and television rights have a fluctuating market, there may come a time when the bond may be too much or too small. So we in the Board felt it was not the proper way to regulate the market.

Q. For the benefit of the public, which seems at a loss about how all these have affected the video rental business, please tell us how the whole system works. As far as the public is concerned, all they see is that all of a sudden most of the video rental shelves are almost empty since you started implementing your new regulations.

A. As far as the trade is concerned, it is very simple. We give them an inventory sticker, the VRB logo. The store just puts the title of the film on the cartridge then attaches this logo. That's it. Besides this, we have what you call the VRB rating label. After May 26, all tapes are supposed to be reviewed and classified. The VRB rating labels are *G* ("general patronage"), *P* ("parental guidance"), and an *R* ("restricted"). The logo that carries the rating label is the logo of the reproducer. In this trade there are two types of merchants involved: the retailers, who deal with the home viewers, and the reproducers, who engage the source for the material and distribute the materials to the retailers.

We talked to the retailers and told them you need an inventory sticker at magkaroon kayo ng VRB label and you need a license, plus you must display the VRB seal outside your establishments.

Q. And how about the reproducers?

A. We give them labels for anything they apply for reproduction. But we ask them to tell who the owners of the films are and whether they have a rightful claim to it. We assign a person to watch the reproduction of a particular tape. We have asked the reproducers to form an association so that they could formulate their own rules and regulations which they would then submit to us for consideration. We want them to impose self-regulation.

Q. What is the arrangement now between the video industry and the movie industry or movie importers?

A. We have asked everyone concerned to iron out all problems in the spirit of partnership. One of our suggestions is for them to form a corporation where they will all work hand in hand in the marketing of videotapes. The proposal from the video people is an offer to film distributors of royalties from between P10,000 to P30,000. They are also offering a protection period of three months after commercial release of film—that is, the film is protected from duplication during that period. The motion picture people would probably want to lengthen this period to four to five months.

Q. Then how did all these rules and regulations result in higher tape rentals for the consumers?

A. Because before nobody paid a royalty. What the consumer will get in exchange is better-quality tapes.

Q. Are you saying that you guarantee that all tapes from now on will be high-quality reproductions?

A. All I can say is that we are imposing ten rules for reproducers to follow and one of them is for them to hire technical people to supervise the reproduction of tapes to ensure that the end-product is acceptable.

Q. What could be the average price increase of video rentals?

A. The average would be a 25 to 50 percent increase in rental prices. The video retailers are now required to pay an excise tax of P5 per tape per year. They are also required now to pay a 30 percent amusement tax.

Q. Isn't the consumer being penalized because government is simply trying to protect the interests of a small lobby group, which is the motion picture producers and importers, who, by the way, are also guilty of a lot of crimes?

A. The contention of the movie industry is that they are an industry while the other one, the video industry, is a racket. You also have to look at it this way, without the motion picture industry there will be no video industry. Plus the government had to intervene because the atmosphere between the video and the movie industries was getting more and more volatile by the day.

4.D: The Uses of Video for Rural Development in Nepal

by Subhadra Belbase

There have been attempts to use VCRs to promote specific aspects of development in the Third World. However, little about these efforts has appeared in print. In the following, Subhadra Belbase discusses how videotapes have been, or could be, used to foster human development. Belbase is with the Worldview International Foundation Media Center in Nepal. The paper was presented in July 1985 at the Development Video Exploration Workshop in Nairobi, Kenya.

The purpose of this paper is to describe some of the ways in which video is being used for developmental purposes in Nepal. I will first describe the variety of ways in which some videos have been used, and report the findings of a survey conducted to study the effectiveness of our programmes. In so doing, I will attempt to highlight some of the advantages of video, and also some of the problems we face while working with this communication medium. Having done this, I will focus on some of the insights gained, and stress the need for a "fresh approach" in the use of video for developmental purposes.

HOW HAS VIDEO BEEN USED THUS FAR?

Video has been used in a variety of ways for developmental purposes by Worldview International Foundation (WIF). I will briefly introduce some of these videos and describe the context in which each was used.

To bring messages from the grass roots to decision makers. "Bartalaap" ("Dialogue") was one such video. Eleven women's projects were evaluated, and on the basis of some general findings common to most of the projects, my collegue and I developed a script. However, when we started to shoot the video which was going to be viewed by decision makers and sponsors, we found that the women who were beneficiaries in the projects had much to say to policy makers. They wanted to be heard. So we threw the script aside and just documented what the women and men involved in the project expressed. The women were not afraid to speak out. They were courageous and knew exactly what they wanted.

This video was later screened at a workshop where some of the rural women, sponsors, and decision makers participated. When the decision makers saw the video, some of them did not like the criticisms made about the projects.

But the rural women seemed to gain courage from the video. They had remained rather quiet prior to the screening, but when they realized how vocal and critical they had been before the camera, they once again started speaking out. On the last day of the workshop, one woman representing the rest said, "Keeping in view the weaknesses in the programmes already implemented so far and the unnecessary difficulties the rural people have had to face, we would like the planners not to repeat the same mistakes."

Did the video change decision makers? We do not know for sure, but it has been called "an eye-operner," and the video, along with the workshop and evaluation report, did make policy makers and sponsors better understand the problems that rural women were facing in these projects which had faulty designs.

Another video that brings messages from rural people to those higher up is "Towards Self-sufficiency." Her Majesty the Queen of Nepal viewed the video where those living on charity were expressing the need to learn skills that would enable them to earn an independent living. They did not want to depend on charity. These people had never met Her Majesty, but video was a medium by which they did reach her.

Video as a training tool. Video has also been used to train health workers for rural work. It is very apparent that when people watch themselves interacting with others, they do discover that the way they talk and behave, body language, and tone of voice, and so on turns people off. Self-awareness creeps in as one watches oneself in a mirror as it were. During such training, I have seen overnight changes in people.

Taking schools and hospitals to the villages through video.   "How many of us go to hospitals to learn what you taught us today? We should be grateful to you for teaching us." This was said by a rural woman whose grandson had recently suffered a bout of diarrhoea. The same woman further explained that when she had taken her grandson to the hospital the nurses there had taught her about ORS [Oral Rehydration Solution], but she had been so worried and concerned about the child that she had not been able to take in any information. But watching the videos in a friendly place where she had no anxieties she was able to learn much.

These educational videos have dealt with information on health and nutrition. Although most are for rural audiences, one, "Save Your Children," is aimed at urban audiences in Kathmandu. The Solid Waste Management Project (SWMP) is using these videos to motivate women to keep courtyards clean. Group discussion after the video is a vital part of the packet.

According to one writeup by SWMP, "Despite the crowd (200 people), it appeared as if the whole film was understood. Most of the questions were replied to by women and children." During a group discussion after "Save Your Children," the women pointed out that they themselves could keep their courtyards clean, and that not enough waste drums were available. Also, overloaded tractors left loose garbage around the neighbourhood. In other words, the video has helped project people understand their beneficiaries better. A long-term use of video is planned for the project.

Advantages of Video

Our series on diarrhoea was shown to over 200 rural people within one hour. I do not think one person or flipcharts or radio could have done this. People were not only listening, and unless they can read instructions on packets or flipcharts the pictures are beyond most people's frame of reference. Hence, audiovisuals like video, as shown from our study, are powerful indeed.

People in Nepal have not yet been saturated with television; hence as one colleague put it, "Just as they learn songs from Hindi movies, little children can start singing simple ditties on health, nutrition, and so on. This is possible, as we have witnessed in our work with children. They pick up words and messages, compose tunes, and start singing them as they leave the video screening.

Disadvantages

This is not to say that we have not encountered problems. Our problems have been purely mechanical. Heat and dust, spare parts, maintenance, and, of course, most villages do not have electricity.

What we need are portable solar-charged recorders and players, accessibility to spare parts, and training in maintenance. Also, it would be wonderful if third World villages were kept in mind when manufacturers designed their products. But of course, videos for rural audiences and developmental communication are sponsored by First World nations—we cannot afford them.

VIDEO SURVEY: DO RURAL PEOPLE LEARN FROM VIDEO?

Background

Some communicators believe that people in rural Nepal will not be able to comprehend moving pictures; hence an expensive technology like video should not be used for rural developmental purposes. Worldview International Foundation had been using video for rural audiences, and seen that rural audiences had not only enjoyed video, but also understood the contents. However, a systematic study to substantiate this finding had not been attempted.

Therefore, on June 3, 1985, WIF went to a rural area called Dhapakhel, a place close to Kathmandu, consisting of a predominantely agricultural Newar community, where the women visit Kathmandu only when there is a severe crisis. A majority of these women are unschooled and cannot read or white. A month earlier we had visited this community and shot "Diarrhoea: How and Why?"

Purpose

The purpose of this study was to find out if rural "illiterate" women gained information from video.

Methodology

A total of 8 unschooled women were selected from Dhapakhel to answer questions on Oral Rehydration Solution (ORS) and nutrition during diarrhoea. Four (4) people answered each questionnaire. One of the subjects, ORS, had already been introduced during the shooting of "Diarrhoea: How and Why?" However, we had not demonstrated how to prepare the solution, but we had verbally mentioned and explained it with the help of flipcharts. The other video "Nutrition During Diarrhoea," had not been introduced.

Our aim was to find out if people remembered concepts used while shooting, and whether there was a difference in grasping concepts introduced before, and not introduced previously.

The questionnaire was administered before and after watching the videos to find out how much information they had before they watched the video and to find if they had gained information after viewing the video.

Findings

ORS

Question: *Should children with diarrhoea be given water or not?*

Responses	Before	After
Yes	4	4
No	—	—

Question: *Do you know about ORS (Aushadi Pani or medicine water)?*

Response	Before	After
Yes	4	4
No	—	—

Question: *Describe how to prepare ORS (amounts of ingredients)*

Responses	Before	After
Yes	3	4
No	1	—

Interpretation

Only one of the women questioned had not taken part in the shooting. This was the woman who could not name the correct ingredients of salt, sugar, and water needed to prepare the ORS.

However, those who had participated in the shooting a month earlier still remembered the details (i.e., half a litre of water, a scoop of sugar, and a pinch of salt).

On probing the woman who had not participated in the shooting, it was discovered that the other women had told her that a child with diarrhoea should be given water.

Conclusion

From this study, we can conclude that those women who participated in the shooting learned during the shooting. Thus if the community participlates, the shooting in itself is educational. Also, the message does spread through interpersonal channels, as is testified by the woman who had heard from the others. A month had elapsed since the shooting took place, yet the women remembered minute details. Hence when people are made to discuss (as we did during the shooting), they do recall.

Women did understand the video, as is verified by the one woman who was able to tell us the ingredients and amounts needed to prepare ORS after the screening, but not before.

Nutrition

Question: *Should a child with diarrhoea be given food?*

Responses	Before	After	Screening
Yes	1	4	
No	3	—	

Question: *Do you know what sarbottam peto is?*

Responses	Before	After
Yes	1	4
No	3	—

Question: *How do you prepare sarbottam peto?*

Responses	Before	After
Correct	1	3
Partially correct	—	1
Incorrect	—	
Don't know	3	

Question: *What happens when a child with diarrhoea is not given any food?*

Responses	Before	After
Susceptible to diseases	—	4
Don't know	3	—
Wrong answer	1	—

Question: *What kinds of diseases?*

Responses	Before	After
all correct	—	1
(5 diseases listed)		
Partially correct	—	1
Cannot recall	—	2

Interpretation and Conclusions

From this study it can be concluded that "illiterate" women who have never seen moving pictures before can not only understand visuals, but can effectively and successfully learn from video.

As can be clearly seen from the findings, only 3 of the 4 women (75 percent) had heard of sarbottam peto, but after the showing all were able to describe how to prepare the cereal for children.

None of the women knew that a child with diarrhoea, if not given food, can catch diseases; but after the screening all—i.e., 100 percent—knew that a child with diarrhoea can catch diseases if not fed. The names of diseases could only be cited correctly by one, another person named two diseases, while the rest (50 percent) could not cite the names. These words and concepts were new, and so it was difficult to recall details.

However, it is interesting that one woman sat for the second time to watch and, on being asked the names of diseases, was able to name them. Hence we can conclude that difficult concepts should be repeated if people are to understand.

INSIGHTS GAINED

And this brings me to some very important points such as scripts, the method of production, realistic depiction, frame of reference, and the potential use of video for developmental purposes.

Scripts and the need for a fresh approach.   I strongly believe that developmental videos do not need strict scripts written by producers, directors, or scriptwriters. Personally I feel that such scripts are needed when the writer has a definite message and wishes the video to convey this message. Such a script might or might not be based on rural needs and aspirations. However, the director needs actors, rural or urban, to perform, or to deliver his dialogue. Words are put into people's mouths, regardless of what people might be wanting to express.

Let me provide an example to clarify my point. A project on family planning might think they are doing the best work in the world, and so decide to portray the success of their project. So they write up a script, and ask rural people, men and women, to say that the family planning and sterilisation project has benefitted them. They have a small, happy family, their children are going to school, their savings have increased, and so on. The reality might be very different indeed. Family planning might have had nothing to do with their increase in savings. And these people might be mourning the death of a baby and wishing they bad not been sterilised. Regardless of these realities, so-called "developmental videos" are produced.

In my opinion, these are not representative of people's thoughts and aspirations. They represent what project people might think about their project. Also, this style represents film making.

In my experience, and according to people who have been using video for developmental purposes, we do need a concept or a broad outline. For example, are we going to make a video on health or some project in that area? We do need to decide the kinds of people we will talk to—women, children, old people (i.e., those involved in that project). However, we must let people decide what they want to say. Just as a good journalist should be impartial, so must the video producer; and just as a good researcher should cover all shades of opinions, so must the video crew. (Similar thoughts and ideas were expressed by Tony Williamson, currently Director of the Don Snowden Centre, Memorial University, Canada, and Dembele Sata Djire Chief, Division for the Promotion of Women, DNAFLA, Republic of Mali, at workshop, "Methods and Media in Community Participation," Labrador, September 28, to October 7, 1984. See also her article, "the use of video in the village," reprinted in Appendix 1.A.")

Mode of production.   A look at the videos I have discussed will demonstrate that we have not hired actors, or put words into the mouths of rural people. Neither did we sit in a studio and produce materials, hoping that rural folk would understand them. What we have done is, let people discuss spontaneously their thoughts, viewpoints, problems, and suggestions on the topic.

Neither have topics been irrelevant to the lives of people. Issues pertinent to the lives of people have been discussed and shot. And the shooting has taken place in rural people's localities. They speak their own thoughts in their own language in familiar surroundings; hence, visuals, sounds, and "characters" are within other rural people's frame of reference. Also there is no need, then, for "simple" dialogue for rural people. They themselves are using everyday speech, idioms, and so forth. Also, then, there is no need for this oft-cited "pretesting" because people themselves are participating. Only when experts in offices produce scripts, visuals, and so on, there is the need to pretest.

Take, for example, our series on diarrhoea. Two of them, "Nutrition" and "Bottle-feeding," are filmed at Kanti Hospital, in Kathmandu. However, our topics are relevant to villagers; and since they cannot go to hospitals to learn, the information from hospitals must be taken to them.

The Potential of Video

Video helps in building confidence.   People must speak for themselves, once they learn to express their feelings—even if it is on tape, they gain confidence. Very often I have come across rural folk—especially women—who tell me that they don't know anything. I tell them that they do know their own village better than I do. She knows something, and I know something. Within two to three minutes they speak, and soon realize that they can express themselves. The light in their eyes, the shy smile, after the "task" is a new confidence. Such confidence is needed—it is a stepping stone to attempt more complex skills.

The need for mobile units and long-term projects.   After having witnessed the effectiveness of video in rural communities, I have come to believe that pertinent information can successfully be taken to villages, and information from villages can be brought to the centres. This can mean that mobile units make periodic visits to rural areas to impart information on hygiene, ORS, simple message on relevant issues.

This does not in any way mean that we should have one-shot exercises in rural communities. Long-term sustained effort in the use of video for motivation, training and education is a must if we are to have a lasting impact.

Of course, video has often been criticized as an expensive medium. But are a few mobile units covering the country any more expensive than the amount of money that has been spent in writing reports, and preparing slides, flipcharts, and brochures? And if one considers that video can reach a far greater number of people within a much shorter time span, and people do get the message immediately, then the cost-effectiveness increases.

Then again, it has been said that educational television reaches only about 1 percent of Third World populations. This means that we need to make a greater effort to reach rural areas and video does have the potential to meet diverse rural needs. Video definitely has the potential, unlike TV, to be viewed when the audience is ready, because a centralized unit does not control screening time. Here again I stress the need for mobile units to cover rural areas. I do not think we have fully exploited the potential of video. It is a new medium, and we need to be innovative and open while utilising video for developmental purposes.

4.E: Video Piracy: The Other Side! (Sri Lanka)

Sri Lanka is a country with moderate per capita income and a relatively high VCR penetration rate. As in the Philippines and South Korea, many citizens in Sri Lanka work in the Arabian Gulf states, and most bring home at least one VCR. Many people on the island make a living by operating tape rental businesses. In this article, which appeared in the November 27, 1984, Colombo *Daily News*, the writer examines some of the economic aspects of home video recorders on this island nation.

With the introduction of Govt's open economic policy, video cassette recorders were introduced to the people of this country on a commercial scale in 1978.

Expensive at first, later models came out considerably cheaper and today a cheap video player can be purchased for around Rs. 8,500/-. *At a rough guess there must be some 25,000 video machines in the country today and more than 70 percent of them belong to middle and lower middle class homes.* Most persons employed in the Middle East send or bring a VCR into the country.

At the early stages the need for viewing material brought VCR owners together and groups of them formed private clubs with a view to pooling and exchanging films which each of them had procured from various sources abroad.

Over the years these private clubs grew in size and some became full-scale commercial enterprises in the business of hiring out video films. Today these clubs and lending libraries hire out video films to their members only, at Rs. 15 per film, for the private home use of such members.

Within the past 12 months or so, a foreign organization called the Motion Picture Association of America and more recently a new local enterprise have launched an intensive attack through the media, against what they call "video pirates," meaning obviously the numerous private clubs and lending libraries which engage in the business of hiring video films.

The term "piracy" in the video business is commonly applied to the unauthorized duplication or transfer of original copyrighted material on to video cassette for the purpose of sale. This does not happen in this country as original copyrighted works are not available here.

Video "piracy" is a huge business in Dubai, Oman, Hongkong, Singapore and the U.K. and that too after spending colossal sums of money.

Their failure to curb this activity in the other countries mentioned is because the laws in those countries are designed to protect their own businessmen rather than to assist the MPAA, which is a colossal and immensely wealthy organization. The loss, if any, to this Organization from the video business in Sri Lanka would be less than a drop in the Ocean.

Every video club and library in Sri Lanka obtains its films from one or more of the countries mentioned above, and mostly through people who travel abroad and those returning home from the Middle East.

Either way some form of duty is paid on the films before they are brought into the country. The net result therefore is that the video clubs and libraries are able to obtain video films at very reasonable prices, usually around Rs. 35/- or Rs. 400/- each, thereby enabling them to hire out such films to their members at Rs. 15/- per film.

The greatest beneficiary of this business is therefore the consumer (VCR owner) who is able to enjoy the film of his choice, with his family within the confines of his home at a price which he can afford.

If the MPAA has its way the middle class video owner will be deprived of this right. He will instead have to pay at least 5 times more for a video film and he will not have a wide choice of films but rather a very limited range, most of which would be very old and shown in local cinemas years earlier.

No existing video club or library in the country has the capability to pay prices quoted for copyrighted films as they operate on the slimmest of margins through a rental process which yields Rs. 15/- per film rented.

The only alternative for them would be to close down, since legal action has been respectedly threatened. If they close down a monopolist in the video rental business in Sri Lanka will emerge.

Such a situation would be to the total detriment of the consumer and the State, for notwithstanding the legal implications a monopoly of this nature would be repugnant to the fundamental principles of free private enterprise and an open economy, which are the avowed policies of our government.

Quite apart from the middle class consumer, consider the position of the State in this matter. As things stand most of the video films available in the country have been purchased out of earnings abroad or from legitimate allowances obtained for travel abroad or have been sent as gifts from friends and relations overseas, so that the foreign exchange outgoing is the barest possible minimum, in fact negligible.

Should the system change in order to foster the interests of the MPAA, the drain on the country's foreign exchange reserves would be considerable.

There must be at present, a minimum of 250 and a maximum of 500 video clubs and libraries in the country, a lot of them situated in and around Colombo and Kandy.

These clubs and libraries have anything between 250 and 4,000 video cassettes each on their shelves. At a rough estimate therefore there may be around 750,000 video cassettes available for hire.

Should even half this quantity be imported from members of the MPAA on the terms offered by their local representative, i.e., approx. U.S.$45 each, the cost to the country would be around U.S. $16,875,000, or approx. Rs. 455,625,000 at the current rate of exchange. The cost is undoubtedly staggering.

Today, after years of hard work and sheer business ingenuity the video clubs and libraries have succeded in bringing a service to the consumer, which is just short of a miracle. What was and still is regarded as a luxury in many countries of the world, has been brought well within the reach of the middle and lower middle classes of Sri Lanka.

For a mere Rs. 15/- these categories of consumers can rent and view in the privacy of their homes, film which a more affluent person in the developed countries would be watching at about the same period.

This was not even a dream a few years ago. This undreamed of facility is, however, in imminent danger of being snatched away from those classes of people who need it most of all since their income just would not permit more costly forms of entertainment.

Should the MPAA and its local agent in their endeavours, the pleasure of watching a video film will most definitely become the privilege of the affluent only since the middle and lower middle classes will never be able to pay the exorbitant rental that will have to be charged by any club or library that can afford to purchase the copyrighted films offered.

Most of the reputed video clubs and libraries in the country are registered businesses and pay their income tax and BTT, although spokesmen for the MPAA and others have alleged otherwise in virulent attacks through the press.

At a rough estimation the state coffers should benefit by about Rs. 25 million through taxation of video clubs and libraries. These clubs and libraries also provide employment to quite a few men and women, and their continued existence is most desirable as they perform an invaluable service to society.

Notwithstanding the legal implications, the question here is whether a giant multinational organization operating thousands of miles away and from whom the State obtains no benefit whatsoever, will be permitted to squeeze out of existence several hundred small Sri Lankan businessmen to the absolute detriment of thousands of middle and lower middle class working people in the country.

CHAPTER 5

Videocassette Recorders in Latin America

Latin America is better defined by culture, language, and history than by geography. It includes most of the Americas, the countries in the Western Hemisphere south of the United States. "Latin" America specifically indicates those countries in which most of the population speaks either Spanish or Portuguese, and which trace significant roots back to Iberia, either Spain or Portugal. This chapter does not include the English- or French-speaking Caribbean, nor Guyana, Surinam, or French Guiana in South America.

Latin America does include Belize because of its mixture of English and Spanish traditions. It includes Spanish-speaking Mexico, Guatemala, Honduras, El Salvador, Nicaragua, Costa Rica, Cuba, the Dominican Republic, Colombia, Venezuela, Ecuador, Peru, Bolivia, Chile, Argentina, Uruguay, and Paraguay, which, as of 1983, accounted for 220 million people, as well as Portuguese-speaking Brazil, which accounted for 129.7 million (World Bank, 1985).

Certain commonalities in Iberian culture were transplanted to most Latin American countries during their colonization by Spain and Portugal. That colonization lasted from about 1500 to the mid-to-late 1800s, longer than the period of colonization of most African or Asian countries. Despite the fact that independence came earlier than to most of the Asian and African countries, which became independent after World War II, most Latin American countries are more deeply marked by the cultures of their colonizers. Indigenous Amerindian cultures persist among sizable numbers of people only in a few Andean and Central American countries. In most countries, the populations are either predominantly European in ancestry, as in Argentina, Uruguay, and Chile; *mestizo*, or mixed Indian-European, as in Mexico, Central America, and the Andean countries; or mixed European and African, with traces of Indian, as in Brazil, Cuba, and the Dominican Republic.

The cultures of Latin America reflect this mixture. All but minorities speak European languages. Iberian literature and Latin American literature in Spanish and Portuguese are predominant. Legal traditions essentially come from Iberia, although the constitutions tend to resemble that of the United States. A related Iberian tradition favors private enterprise, but under strong control of the state, which itself is heavily influenced or even controlled by key economic groups or families. The military and the Roman Catholic Church have also traditionally had a privileged position and wielded influence in governments.

After independence, however, events have drawn Latin American nations further apart. Different patterns of slave importation and European in-migration have markedly differentiated the current ethnic makeup of the countries. In Argentina, Chile, and Uruguay, Italian, Lebanese, and German immigrants have strongly influenced national development. As Spanish and Portugese economic and political influence diminished, the United States, Great Britain, and France exerted considerable influence on particular countries.

Overall, however, the United States has become the dominant partner and model over the last 100 years. The Latin American countries have copied its institutions; trade and foreign investment from the United States have increased; and the United States has asserted political primacy over the region, dating from the Monroe Doctrine, which announced United States intentions to prevent intervention by non-hemispheric powers and effectively limited potential competition to United States influence. Currently, their economic and political dependency on the United States is a source of concern to most Latin American countries. A popular Mexican saying, attributed to Mexican President Porfirio Díaz, is, "Poor Mexico, so far from God and so close to the United States."

As the population split between Portuguese-speaking Brazil and Spanish-speaking Latin America indicates, strong language and cultural divisions, as well as commonalities, exist. In fact, conversational Spanish itself differs so widely between a number of countries and regions as to be scarcely mutually comprehensible. Still, culture in the form of music, novels, *fotonovelas* (comiclike photo novels), films, and *telenovelas* (television soap operas/serials) flows back and forth in translation and "neutral" Spanish to give the region a cultural identity. Certain shared experiences, such as relatively early independence from colonial powers (in the mid- to late nineteenth century) and subsequent economic and political dependence on the United States and, to a lesser degree, Europe, also tie Latin America together and differentiate it from other Third World nations.

This chapter will focus primarily on a few countries in which VCR diffusion is quite widespread, notably Colombia and Venezuela, as well as on a few countries, such as Argentina and Brazil, in which VCR use seems to be picking up. For contrast, it will look as well at places where VCRs are either uncommon, such as the Dominican Republic, or used in small-scale ways for revolutionary or alternative purposes, such as El Salvador or Nicaragua. It will also examine the key role of a couple of countries, particularly Panama and Mexico, in VCR hardware and/or software distribution.

COMMON ELEMENTS OF LATIN AMERICAN BROADCASTING

After a century of diverging political systems, varied economic successes and failures, and emerging ethnic differences in Latin America, it sometimes seems that, across the region, broadcasting has shown more commonalities than many other areas of culture and activity. There are a number of things which nearly all Latin American broadcasting systems have in common: predominantly private ownership of broadcasting, principal financial support by advertising, governmental controls that vary but tend to be less overt than in most Third World countries (except in content censorship), relatively high diffusion and dispersion of radios and television sets, and emphasis on entertainment in radio and television.

A number of these commonalities come from the North American model of privately owned commercial broadcasting that nearly all Latin American countries adopted, in contrast to most other Third World nations (Katz & Wedell, 1977). Several studies have shown how United States equipment suppliers, program suppliers, and advertisers were very influential in creating private advertiser-supported, entertainment-oriented broadcasting systems in Latin America (Road, 1976; Tunstall, 1977; Fejes, 1980). In some cases, United States advertiser influence seems to have tilted countries such as Colombia away from initial decisions to pursue educationally oriented television toward systems that allowed for more advertising and entertainment (Fox de Cardona, 1975). In Cuba and Nicaragua, socialist political revolutions have led to the nationalization of previously private commercial broadcasting systems.

However, not all, and perhaps not even the most important, influences toward commercialization of Latin American broadcasting came from the United States. Nearly all Latin American countries show a pattern of powerful industrial families or groups taking the initiative to begin broadcast companies. Sometimes government leaders simply permitted this; others, such as former President Miguel Alemán in Mexico, emerged with an equity interest in television systems themselves (Mahan, 1985). Still other leaders, such as Vargas in Brazil, worked with broadcast industrialists, such as television founder Assis Chateaubriand, to arrange favorable coverage for themselves and their programs (Straubhaar, 1981). To a large degree, this seems to be part of a pattern of government-sanctioned private monopolies or privileged oligopolies in various parts of the economy that can be traced back to Spanish and Portuguese roots (Wiarda, 1981).

In their survey of Third World broadcasters, Katz and Wedell (1977) classify most Latin American systems as mixed, with both private and government ownership, advertising and public support. However, since the public systems tend to be relatively small and underfunded, except in Cuba and Nicaragua, such classifications can be deceptive. The overall nature of broadcasting in Latin America is clearly private and commercial. Even Colombia, where the government owns broadcast facilities and uses them for some daytime educational programs, leases its television channels out to commercial interests in prime time (Ferreira & Duke, forthcoming). In Chile, where universities own several major television channels, those channels have come to rely on advertising and operate in a commercial manner (McAnany, 1987).

Latin American governments originally tended to take a laissez-faire, or hands-off, approach to broadcasting, except to regulate frequency or channel allocations and censor material for political or "moral" reasons. As de Camargo (1975) notes for Brazil, Latin American governments tended to let private groups initiate and invest in broadcasting, so that scarce state revenues could be shifted elsewhere. This occurred also because Latin American governments and businesses tended to respond to United States influences and the United States model of private ownership. For many years, media-state relations in most Latin American countries seemed to fit the "authoritarian" model reasonably well, engaging in periodic or even regular censorship, but not otherwise owning or controlling broadcasting. However, as we shall see below, this generalization is becoming much less accurate.

Given its private advertising-supported structure, it is not surprising that television in Latin America has been entertainment oriented. A study in Mexico showed that between 1953 and 1976, an average of 81 percent of television programming emphasized entertainment (Rota, 1985, p. 204). A study in Brazil showed that between 1963 and 1977, entertainment averaged 80 percent of broadcast time.

Furthermore, the Brazil study showed that an even higher average (86 percent) of viewing time was spent watching entertainment (Straubhaar, 1981; 1984, pp. 40–43). The predominance of entertainment in Latin American broadcast television means that there is not as much unfulfilled demand for entertainment in most Latin American countries as there is in some other regions of the world, where entertainment is less dominant in broadcast content. That seems, in a number of cases, to lower the demand for VCRs, as we will see below.

News programs are also prominent in most Latin American television systems, because they are popular. Research in Brazil (da Via, 1977; Lins da Silva, 1985) shows that working-class viewers, as well as elites, are regular news watchers and list it high in their preferences. Hence, many of the periodic struggles between government censors and broadcasters hinge on television news, which is a particularly important news source for the public because radio networks are often loosely organized. At least in Mexico (USIA, 1981) and Brazil (USIA, 1982), where the press is regionalized, national television network news is the only common news source for both the elites and the general public. Festa and Santoro (1987) note that, in Brazil, self-censorship by television networks has to a large degree replaced government censorship, which has greatly declined during the transfer of power from military to civilian regimes. Even under civilian rule, however, a number of points of view are excluded from coverage in broadcast news. This continuing exclusion feeds the use of video as an alternative means for both media professionals and community groups to put together "news," documentaries, and dramatic programs that could not or would not be broadcast.

Some governments, notably that in Colombia, and to a lesser degree, those in Brazil, Mexico, and elsewhere, require broadcasters to carry certain amounts of educational programming. Some of this is produced by government institutes, as in Mexico and Colombia; some by commercial broadcasters, in cooperation with government and private foundations, as in Brazil. What is most distinctive in Latin American television education is the tendency to integrate development messages and broadly educational themes into entertainment programs, particularly prime-time serials or telenovelas. It began when governments and development groups observed the social effects of some telenovelas. For example, in the Peruvian program "Simplemente Maria," a poor girl moves to the city, saves her earnings as a maid, buys a sewing machine, and becomes a seamstress. In a clear imitation effect, all the sewing machines in Lima were immediately bought out.

Governments have since tried to incorporate "prosocial" themes in a number of programs, which have been noteworthy in some development campaigns. Formal or semiformal education, such as literacy and math, has been taught more often with radio than with television in Latin America, with extremely successful operations sponsored by the Roman Catholic Church, the United States Agency for International Development, and national governments in Colombia, Nicaragua, and the Dominican Republic (Jamison & McAnany, 1978).

Given the predominance of entertainment in Latin American television, a major issue concerns the origins of that entertainment. Live local production in the 1950s gave way to a wave of imported United States series, cartoons, and feature films in the 1960s as kinescopes, videotape, and microwave technologies permitted the use of prerecorded (imported) programs and multicity networking. In the 1970s, local programming, particularly marathon variety shows, music programs, and *telenovelas*, emerged strongly again in Argentina, Brazil, Venezuela, and Mexico. The 1980s have shown a further decline in imported programs in some countries, notably Brazil, Chile, and Venezuela, and a tendency in smaller Latin American countries to create more music, variety, and interview programs, and to substitute regional productions from Brazil or Mexico for programs from the United States (Varis, 1984; Straubhaar, forthcoming).

This tendency to regionalize importation of television programs seems to affect VCR programming and diffusion directly, since most videotapes circulated in Latin America still contain North American movies. What are the implications for the demand for such movies, if television fare is regionalized? VCR use seems to have gone up in Venezuela, at least, to accommodate the lingering interest of some in watching imported films, which are now less available on broadcast television.

Given that creation of television channels in Latin America has largely been left to private initiative, under varying degrees of regulation, it is not surprising that most Latin American countries have several channels available throughout most of the populated areas. A safe generalization is that there are usually about as many channels as advertising will support. Those countries that have restricted the number of channels, most notably Colombia and Venezuela, seem to have had the fastest diffusion of VCRs. This connection between television diversity, at least the number of channels, and VCR diffusion will be explored further below.

Beyond the availability of signals in most populated areas, television sets and television viewing are also much more widespread in most Latin American countries than in most other Third World nations, except perhaps those of the Arab world and some in East Asia. In Argentina and Uruguay, over 90 percent of the population has a television set at home. In Brazil and Mexico, 70 to 80 percent regularly see television; that proportion will probably expand further as satellite delivery of signals to remote repeater stations expands. That means that more homes have television sets to connect to VCRs, but it may also mean that VCRs will have to compete with regular signals. In Peru, Bolivia, and some Caribbean nations, for example, VCRs now play a major role in introducing "television" to remote areas (Mayo et al., 1987).

ECONOMICS, BROADCASTING, AND VCR OWNERSHIP

Latin American specialists themselves argue about the distinctiveness or commonality of the various cultures and political systems in the region. While the economies vary more obviously in their size, diversity, and external importance, there also exist differences of opinion about which economies perform better in terms of growth, income distribution, control of inflation, employment, and other dimensions. Major divergences in governmental economic and media policies have also begun to affect the dispersion of VCRs in several countries. The relative homogeneity of various cultures, the varied strength of the media industries that feed television, and the diversity of television offerings are also factors that differentiate countries for television development and VCR diffusion and impact.

First, the overall wealth and the income distribution of Latin American countries vary considerably. Brazil is currently the ninth largest economy in the world and Mexico is fourteenth, in terms of overall Gross National Product (Kurian, 1984, p. 95). Still, some Latin American countries like Bolivia and Guatemala are among the poorest and most limited economies in the world, exporting only one or two primary products, such as tin or bananas. Crude economic indicators such as GNP per capita do make useful distinctions among these countries. Relatively higher average income among nations does, in fact, correlate positively with national VCR penetration (Straubhaar & Lin, in press). Relatively higher wealth permits more people to afford a luxury like a VCR. That poorer countries such as Bolivia, Paraguay, the Dominican Republic, and those in Central America have fewer VCRs per capita is no surprise.

However, lower overall income levels in some countries like Bolivia and Paraguay, in parts of Peru, and in much of Central America, have also made the extension of television broadcasts to remote areas difficult. That creates a latent demand for some kind of television, whether by VCR, satellite dish, or other means. When incomes in these areas rise, some individuals acquire VCRs. This seems to be true in rural Bolivia and Peru, where peasants with newly increased income, probably based in the cocaine trade, frequently acquire VCRs and/or satellite dishes for television reception (Mayo, 1987).

Second, while income distribution is a subtler concept and more difficult to measure, it also is important for VCR diffusion. High average income is not as meaningful for predicting purchases as it seems if the distribution of that income is highly skewed toward a small group of wealthy people. At one level, in consumer-oriented economies like those of Argentina, Brazil, or Venezuela, skewed income distribution reduces the number of people who can afford a VCR. At another level, particularly in the poorest countries, the concentration of income does permit a few among the elite to afford a VCR (whereas if income were distributed equally, there would be more radios and small television sets but fewer VCRs).

Third, government policy toward broadcasting, particularly television, has begun to differ greatly among Latin American countries in recent years. Given scarce resources and the willingness of private investors to build and operate broadcast facilities, many governments did, in fact, turn their attention elsewhere for many years. More recently, however, governments in the larger nations have become more involved in their economies and more activist in policy-making; thus some of them have begun to intervene more in broadcasting. Formal intervention tends to be episodic, however, rising with activist governments such as those of Peron in Argentina, Vargas in Brazil, or the military in Peru in 1968.

Informal intervention by Latin American governments in broadcasting has become more common in recent years. In Brazil, the post-1964 military governments subsidized the microwave and satellite facilities that enabled private television to reach all of the country, because that fit military goals for national integration (Mattos, 1984). The Brazilian government also subsidized the buy-out of Time-Life in order to "nationalize" private TV-Globo (which had been started as joint venture with Time-Life) and leaned on the network to give favorable coverage to government development programs (Straubhaar, 1984). In Mexico, the government initially used the private television network, Televisa, to spread general propaganda and even specific messages on things like family planning. For example, several serial programs (*telenovelas*) were specifically created by Televisa in cooperation with government planners to promote birth control. More recently, however, Mexico has started to build up a major government network to compete with Televisa.

There are several implications for VCR acquisition in Latin American governments' changing approaches to broadcasting. If government control of content becomes more prevalent, then VCRs can be a means of adding forbidden content, whether entertainment or information. If government stations succeed in adding more diversity and appealing to at least some new segments of the audience, however, then there is that much less latent demand to be met through VCRs.

Fourth, although Latin American countries are less diverse ethnically and linguistically than India or most African countries, there is an important variation in cultural homogeneity between countries. Some countries, notably Bolivia, Guatemala, Mexico, and Peru, have substantial indigenous populations that do not speak either Spanish or Portuguese, and therefore have less access to the major media systems. Some others, such as Argentina, Brazil, Chile, Uruguay, and Venezuela, have substantial immigrant colonies from Europe or Japan. While most of these immigrants seem to be assimilated into the national culture and language, there are minorities who probably are interested in films or programming from their linguistic or ethnic homelands. These groups seem to have some natural use for VCRs in acquiring foreign-language films and television programs.

Fifth, in some countries, other cultural industries feed material to television. Argentina, Brazil, and Mexico all have music, theater, film, and publishing industries to supply actors, musicians, comedians, directors, scripts, music, graphics, and other aspects of television production. These enable some Latin American television industries to produce more entertainment programming: soap operas (telenovelas), variety shows, music, comedies, dramas, and miniseries. At least in Brazil, however, the growth in the number of professionals and eager amateurs in music, theater, and "television" production has led many of them to work in video, either as an alternative to a regular media job or as a means of gaining experience and visibility. Music video and alternative drama have been particularly prominent.

The ability to produce programming of relatively high quality is reflected in the degree to which Latin American nations import or produce programs. As of 1982–1983, several produced half or more of their own programming: Argentina (60 percent), Brazil (61 percent), Chile (64 percent), Mexico (50 percent), and Venezuela (67 percent). Others produced much less: Ecuador (34 percent) and Peru (30 percent) (Antola & Rogers, 1984; Varis, 1984). The quantity and quality of nationally produced television programming affect VCR diffusion by satisfying audience needs for entertainment, the material may be more appealing by its cultural fit to national audiences than the typical pirated foreign movie fare for VCRs.

Sixth, the relative diversity of television programming, including entertainment, is also critical for the relationship of VCRs and television. In most countries, a lack of diversity is effectively defined by limits placed on kinds of content by broadcasting organizations. In many regions and countries, particularly outside of Latin America, the most critical issue in diversity is entertainment. The broadcast goals of many countries emphasize education, information, or propaganda over entertainment. In those countries, VCRs come to have a major function in supplying entertainment in the form of series, music, comedies, feature films, and so on. This issue seems less important at Latin America, given the entertainment focus of most systems.

However, if one looks more narrowly at kinds of content that broadcasters might exclude, there are many things that Latin American audiences might find missing, depending on the country and government policy in force at the time. Such content includes sexual activity (implicit or explicit), dissident social values or norms, dissident political views, violence, consumption or materialism, and minority languages and cultures.

For example, in Chile, the media censorship and repression of political discussion under the government of General Pinochet keep a great deal out of television news. Chilean censorship also removes from television the expression of alternative social, sexual, and religious values that are only indirectly political. One response has been to create alternative documentaries, interviews, protest music, and other information on VCRs and circulate copies in personal and political group networks (Karen Ranucci, personal communication, October 1986).

Most of the Latin American governments censor from broadcast television pornography and material that offends national morals and customs. However, this is not nearly as major an issue as in the Arab world or even in many other Asian or African countries. In Brazil, for instance, commercials and entertainment programs contain degrees of nudity that would not appear on United States television. Similarly, while some material may be considered overly violent, Latin American standards on television violence tend to be similar to those of the United States. Among these types of content, there are not many— except perhaps pornography and opposition political views—so lacking from broadcast television that viewers are motivated to use VCRs to obtain them.

An equally critical issue in diversity in Latin American broadcasting relates to the specialization or segmentation of programming according to audience interests. In several countries, notably Brazil and Mexico, there has been considerable segmentation of the television audience along age and social class lines. In Mexico, the multichannel private network, Televisa, aims one channel at younger people, another at middle-class viewers, and still another at the broad general audience. In Brazil, different networks target the general audience, the middle and upper classes, and the lower middle and lower classes. In countries where this has not taken place, these audience segments may turn to VCRs in order to obtain programs that fit their interest.

The internationalized middle class and elite in Brazil and Mexico seem to be satisfied by such segmentation, while they are prone to use VCRs in some other countries. This group seems, from available ratings data in Brazil, Mexico, and Venezuela, to be proportionately more interested in imported American feature films and series. Since broadcast television for the general public in most Latin American countries is gradually replacing such imported material with domestic *telenovelas*, variety shows, comedies, and music, as reported above, the middle- and upper-class groups are likely to be somewhat dissatisfied. This seems to be the case in Venezuela, for instance, where of the three channels, none primarily targets the middle and upper classes (interview, Clemente Cohen, Vice-ministro de Información y Turismo, Caracas, Venezuela, June 1985). since this group also has the most purchasing power, it is not surprising that they acquire VCRs and use them primarily for United States feature films and series (see Table 5.3).

In order to assess its effects on VCR diffusion and use, perhaps the simplest measure of diversity is the number of channels of broadcast television available. In both Brazil and Mexico, channel segmentation developed only after a critical number of national networks, usually 3 or 4, developed, and as the logic of commercial competition led some or all of the networks to specialize in order to capture a certain segment of the market, much as commercial AM and FM radio had already done in both the United States and Latin America. A study of eighty-one countries, including most of the Latin American nations, by Straubhaar and Lin (in press) showed that a significant correlation existed between higher VCR penetration and a low number of broadcast or television channels, particularly in Third World or lower-income countries, where time shifting is not a particularly common use of VCRs. Of Latin American countries with relatively high GNP per capita or purchasing power (see Table 5.2), those with more television channels and/or growing use of cable TV (Argentina, Brazil, Costa Rica, the Dominican Republic, and Mexico) have lower VCR penetration rates than those with fewer broadcast or cable channels (Colombia, Venezuela, or Peru).

FILM AUDIENCES

Latin American film audiences have fairly extensive access to a wide variety of North American and European movies by theatrical distribution. The average Latin American sees more films per year (from 0.9 in Bolivia to 4.1 in Mexico and 4 in Colombia) than do people from most of the Third world, where the average is 2.7 for low-income and 1.1 for middle-income nations (UNESCO, 1980; 1981). Cinema admission prices are relatively low throughout the region and theaters are relatively widespread, at least in urban areas. In 1977, Latin America had 22 cinema seats per 1,000 inhabitants while Africa had 4, Asia 8.6, and the Arab states 7.4, compared to 52 in North America and 39 in Europe (McBride Commission, 1980, p. 125). In 1978, there were 2,973 theaters in Brazil, 2,400 in Mexico, 1,004 in Argentina, and 682 in Colombia (Schnitman, 1984, p. 116).

Some Latin American countries have developed productive film industries. In 1981, Argentina produced 24 films per year; Mexico, 65; and Brazil, 80 (Schnitman, 1984, p. 118). However, even these countries cannot produce enough films to satisfy local demand. For example, Brazilian authorities require that 140 days per theater per year be filled by Brazilian productions; to meet that quota, a number of low-budget comedy and pornographic films are produced, as well as a number of relatively high-budget feature films (Johnson & Stam, 1982).

Relatively high cinema attendance in Latin America, contrasted with relatively low levels of cinematic production in all but a few countries, clearly shows that an audience has been created over the years for imported feature films. This audience has been cultivated in part by aggressive marketing by the Motion Picture Export Association of America, its member companies, and European distributors (Guback, Varis, et al., 1982).

Certain genres of imported film have hit a responsive chord in Latin America, creating patterns of taste that have been building since the first export boom from Europe and the United States in the 1910s and 1920s. Action adventure movies have been a particular province of imported films: westerns, spy stories, detective shows, thrillers, war movies, and historial epics. These have been hard for Latin American cinema industries with limited budgets to compete with. Imported musicals, romantic stories, and comedies have also been popular, but the production of competing national equivalents for these genres is somewhat easier for national film industries. In fact, however, a great deal of the talent, production resources, and audience for music, romantic fiction, and comedy has been absorbed by national or regional television productions.

VCRs have provided imported feature films to the Latin American audience. Videocassettes have several advantages over theaters: a wider variety of titles available through rental shops or video clubs; more flexible viewing times; and less need to travel or find a parking space or walk. In some cases, videocassette rentals may be cheaper than cinema admissions, as in some Brazilian cities; but, in general, at U.S.$1–$2, they are at least double the price of cinema admissions in most Latin American countries (see Table 5.1). However, in Latin America, certain cultural factors tend to keep cinema attendance up despite VCR competition. Primary, perhaps, is the desirability for many Latin Americans of going out rather than staying home. This may change with transportation problems, perceived danger of street crime, and cost. Still, there is stronger tradition in Latin America than in some regions, such as the Arab world, of men, women, and adolescents all going out for evening activities.

PIRATED VIDEOTAPES AND VCR DIFFUSION

Building on this potential base in the film audience in Latin America, VCRs have spread in some countries, in part, as a function of the availability of a wide variety of low-cost feature films on tape. Both the variety and low cost are related to the fact that most videotapes in circulation are pirated.

The method of operation and impact of videotape piracy will be discussed below, but a necessary question here is the degree to which piracy has influenced the adoption and use of VCRs in Latin America. Two factors in this question are the degree to which pre-recorded tapes in circulation are pirated, rather than legitimate, copies, and the degree to which VCR use is primarily for playing back prerecorded entertainment—e.g., largely imported feature films.

Table 5.1 shows that, except in Venezuela, most tapes in circulation are, in fact, pirated, and that their cost is quite low. Although it is hard to demonstrate the connection quantitatively, interviews with video rental dealers in several countries indicate that rental prices of pirated material tend to be substantially lower than for legitimate copies, since no royalty payment is made to the holder of the copyright. This lower price brings in a broader potential pool of renters/borrowers. The trade-off for customers is that pirated copies are frequently of lower quality. As seen below, an appeal to an interest in higher-quality

TABLE 5.1   VIDEO TAPE RENTALS, COST AND PIRACY

Country	Tape Rental Cost	Tape Turnover Per Month	% Tapes Pirated	VCR Cost
Argentina	U.S.$2.50	12,000	50%	U.S.$800
Brazil	0.40	50,000	95	500\1,000[a]
Chile	1.20	12,000	100	—
Colombia	1.60	12,000	98	400\800
Peru	1.00	—	100	1,773
Uruguay	2.00	—	100	—
Venezuela	1.00	14,000	25–30	400\690

[a] The first figure is the black market price and the second is the official price, where such a distinction is clear.

Source: NTC/NCT Newsletter (Nuevas tecnologías de comunicación/New communication technologies) *Lima, January 1986, p. 5. Based primarily on 1985 data. VCR cost figures also draw on correspondence by author with various sources.*

copies for viewing has been part of the campaign to promote legitimate copies in Venezuela.

A review by the *NTC/NCT Newsletter* (1986) noted that "the primary use of video recorders in Latin America has been for private domestic viewing of prerecorded tapes." This seems to be especially the case in nations with higher GNPs per capita, such as Venezuela (U.S.$3,840—see Table 5.2), where the video market is 99 percent for private domestic use and only 1 percent for institutional use. In contrast, Peru has a relatively lower GNP (U.S.$1,040) and higher VCR costs because of import restrictions (Table 5.1), so most VCRs seem to be used by institutions rather than private owners (*NTC/NCT Newsletter*, 1986).

From the review of Latin American cinema and film attendance patterns above, it seems that most of the rental or video club tapes available for private playback are imported feature films. This impression is reinforced by a review of the rental catalogs of VCR distributors in Argentina, Brazil, Colombia, the Dominican Republic, and Venezuela. Interviews by the author and colleagues with rental shop owners in these countries confirm that imported feature films have the widest circulation. In São Paulo, Brazil, a newspaper survey of rental shops and clubs showed that, over a six-month period in 1986, a total of 92 percent of the top ten videotape rental titles were imported films (United States, 76 percent; Australian, 10 percent; Argentine, 3 percent; British, 2 percent; and Japanese, 1 percent) (Datafolha/*Folha de São Paulo*, 1986). Data about actual audience behavior is limited but tends to confirm what the pattern of tape availability suggests. A 1987 survey of television households in Santo Domingo, capital of the Dominican Republic, showed that of the 4 percent who had a VCR, two-thirds used them to watch feature films, mostly imported from the United States (Straubhaar & Viscasillas, 1987).

A final point on the relationship of VCR adoption/use to demand for imported films has to do with differences in that demand among the various social classes in Latin American societies. Particularly important is the degree to which the tastes of different classes are met by broadcast television. Different levels of audience segment satisfaction seem to be revealed in the pattern of who has acquired VCRs and what use they have made of them. Acquired primarily by the upper class (see data on Brazil, the Dominican Republic, and Venezuela below), VCRs are used primarily to substitute imported films for what might otherwise be watched on television. The class aspect of VCR usage varies between countries, but case studies reveal major class differences in reaction to the popular culture that is broadcast over television—differences perhaps more important in VCR acquisition and use than those in purchasing power.

Depending on the country, the greater buying power of the upper socioeconomic groups is also reinforced by a greater coopting of upper- and middle-class groups into international patterns of taste, based on contact with foreign corporations, travelers, and films. These groups are also more likely to speak English. These factors make them more inclined to substitute imported films and other imported material for what is broadcast locally. In places such as Venezuela, where increased affluence has brought VCRs within the reach of a broader spectrum of people, the use pattern seems to be somewhat different; here, there is more interest in programs produced nationally or at least dubbed into Spanish. In any case, VCRs have a tendency to fragment the viewing audience and reduce the likelihood that people of various social classes will be exposed to the same cultural productions.

ALTERNATIVE PRODUCTION AND VCR DIFFUSION

Although VCRs have spread initially in countries like Venezuela and Colombia as a means of viewing imported, largely pirated feature films, other patterns are emerging in Brazil, El Salvador, and Nicaragua. There, VCR use seems more linked to group or individual production within these countries. Relatively more cameras are sold along with VCRs, and relatively more VCRs are sold to institutions and groups (Santoro, 1985; *NTC/NCT Newsletter*, 1986).

In Brazil, nationally made how-to-do-it tapes, home recordings, and alternative programming of varying sophistication by unions, feminist groups, ecology groups, political parties, and video artists are all relatively more popular than in many other countries. According to *Veja* ("Videocassete no Brasil," 1986), such alternative uses have contributed greatly to the boom in VCR sales and use in Brazil in 1986, beyond the diffusion of video and VCRs driven by the urge to see more filmed entertainment.

In El Salvador and Nicaragua, video has joined clandestine and semiclandestine radio as a major tool of persuasion by guerrilla groups and governments. Such alternative uses of VCRs and exchanges of the tapes produced are being promoted by the groups mentioned above, the Roman Catholic Church, and others. It is hard to estimate the audience for such VCR productions, but they have already affected the style of television production.

VCR DISTRIBUTION AND SMUGGLING

VCRs have entered Latin America in a number of ways. They have been legally imported, illegally imported, and, in a few places, produced in-country, usually from kits imported from Japan or Europe. Of these three means, illegal imports have probably accounted for the largest number of existing VCRs in most Latin American countries.

Manufacturers' sales data indicate, for instance, that if the total number of VCRs (387,809) shipped to Panama by Japanese manufacturers had stayed in the country, there would be 1.76 VCRs for every television set in Panama (Ganley & Ganley, 1986, p. 25). Instead, penetration is about 10 percent of television households, according to the Motion Picture Export Association's estimate (cited in Ganley & Ganley, 1986, p. 25). Most of the VCRs legally imported into Panama, either through the Canal Zone or the Free Trade Zone, have been passed through and sold to other Latin American countries such as Colombia. Of the 500,000 machines in Colombia in 1985, "most were smuggled in from Panama" (Ganley & Ganley, 1986, p. 25).

Most governments in Latin America have tried to erect trade barriers to the importation of VCRs. For countries with balance of payments problems and large national debts requiring service, VCRs seem extravagant consumer purchases. Brazil has prohibited VCR importation, except through the Manaus Free Zone, and then with heavy tariffs. One effect has been to make a VCR, legally imported or manufactured via joint venture in Manaus, three times as expensive as the world price (*Veja*, 1981, p. 41). Mexico has prohibited the importation of VCRs since 1982. Others have followed suit.

However, import prohibitions and tariff barriers have been largely circumvented by smuggling. A smuggled VCR in Brazil in 1982 cost more than the untaxed world price, but was half the price (Cz$200,000 vs. Cz$390,000) of a legal, domestic model. In 1985, Mexico had about 500,000 VCRs, even though it had only 180,000 when importation was legally stopped in 1982.

There are major smuggling networks for both VCRs and videocassettes through Panama, Miami, New York, Colombia, and Paraguay. For Mexico, most smuggling comes in straight across the United States border. For Central America, Colombia, and Venezuela, most smuggled VCRs or tapes probably come in either directly from Miami or through Panama. For southern South America, particularly Argentina and Brazil, smuggling tends to come from Miami, New York, Paraguay, or a combination of these (*Veja*, 1981, p. 41; Mattelart & Schmucler, 1985, pp. 34–37).

A number of countries have responded by creating "free zones" for importing goods at reduced tax rates to cut down on smuggling. In some cases, these zones also serve as bases for joint ventures between national and multinational firms, which produce products for domestic markets or export. To attract firms to these areas, governments often reduce or waive import duties, import quotas, and a range of national taxes. Panama (Free Port of Colón), Brazil (Manaus), Bolivia (Iriqu), Chile (Iquique), Colombia (Baranquilla), the Dominican Republic (La Romana), and Mexico (the "bracero plan" areas on the United States border) are all major free zones. These are also being developed in Argentina, Barbados, Costa Rica, Ecuador, El Salvador, Haiti, Honduras, Jamaica, Paraguay, Peru, Trinidad, and Uruguay (Mattelart & Schmucler, 1985, pp. 22–32). Among these, only Brazil and Argentina have made serious strides toward replacing illegally imported VCRs with legal, nationally produced ones, by increasing production in-country and cutting taxes and costs.

LATIN AMERICAN VCR OWNERSHIP

Not all countries show much spread or diffusion of VCRs. While income and income distribution divide countries into groups more and less likely to be able to afford VCRs, a given level of income, represented best, if not adequately, by GNP per capita, does not always predict whether a country's population will use its resources for VCRs.

TABLE 5.2 1984–1985 VCR PENETRATION AND GNP PER CAPITA

Country	VCR Penetration in TV Households	Population	GNP Per Capita (U.S.$)
Argentina	1.4%	0.3%	$2,767
Brazil	6.3	1.5	1,880
Chile	2.5	0.5	1,870
Colombia	17.7	1.2	1,430
Dominican Rep.	2.5	0.2	1,370
El Salvador	1.7	0.1	710
Ecuador	10.0	0.4	1,180
Honduras	3.0	0.2	670
Jamaica	7.8	0.1	1,300
Mexico	5.0	0.6	2,240
Panama	10.0[a]	1.0[a]	2,120
Peru	12.1	0.4	1,040
Puerto Rico	14.6	3.8	2,890
Uruguay	7.0	1.1	2,490
Venezuela	31.3	3.2	3,840

[a] Panama figures are an estimate (Ganley & Ganley, 1986, p. 26). Manufacturers' figures are suspect because many VCRs are resold or smuggled through Panama, and sales figures may not represent VCR possession by Panamanians.

Source: VCR data on Argentina, Chile, Colombia, Ecuador, Peru, and Venezuela are taken from NTC/NCT Newsletter 1986. Brazilian data are taken from Veja (December 28, 1986). All other figures are based on data from electronics exporters about VCR sales, compared with estimates of VCR possession made by Motion Picture Export Association and other industry groups.

Table 5.2 shows the penetration by VCRs in the total number of households in various nations, compared with the GNP per capita of those countries.

Despite the greater income in many Latin American countries, VCR spread or penetration is frequently greater elsewhere in Arab or Asian countries. In Latin America, VCR penetration is highest in Venezuela and Colombia.

Venezuela

The home use of videocassette recorders appears to be more widespread in Venezuela (31.3 percent of TV homes) than in any other Latin American country. VCRs diffused rapidly in the late 1970s and early 1980s because many Venezuelans could afford them during that

TABLE 5.3 VCR OWNERSHIP AND SOCIAL CLASS IN BARQUISIMETO, VENEZ., 1985

	Class D (lower)	Class C (lower middle)	Class B (middle)	Class A (upper middle & upper)
Own VCR	0 %	4 %	31 %	65 %
No VCR	100	96	69	35
Total	(176)	(148)	(146)	(88)[a]

[a] Based on the proportions indicated in the last (1980) census, in this sample of 605 adults (16 or over) in the metropolitan Barquisimeto region, Class D contains the lowest (in terms of socioeconomic status) 40 percent of the sample, Class C has 34 percent, class B has 23 percent, and Class A has 3 percent. Social class is measured by income, education, and possessions. Data cited by permission of Fausto Izcaray.

Source: From Some remarks on the agenda for communication research in the 80's: The impact of the new information technologies on Latin American societies by F. Izcaray, 1984, paper presented at the International Association for Mass Communication Research meeting, Paris.

period of relative affluence, and because Venezuelan governments had postponed the introduction of color television until 1980. Before that, Venezuelans who wished to watch color programs from other countries acquired VCRs (Izcaray, 1984).

Overall, about 17 to 18 percent of the households in Venezuela had a VCR, as of 1983–1984 ("La industria de lo audiovisual," 1985). As in most countries, VCR penetration is higher in larger cities: in 1983 it was 30 percent in Caracas; in 1985 it was 11 percent in Barquisimeto (Izcaray, 1985).

Videocassette recorder ownership is also extremely stratified by social class. A 1985 study by Izcaray in Barquisimeto (a medium-sized city in the interior of Venezuela) showed no households with videocassette recorders in the lowest of four social class groups, only 4 percent in the lower middle class, 31 percent in the middle class, and 65 percent in the upper class (see Table 5.3).

A breakdown by education showed less stratification but still a tendency for VCR ownership to be concentrated among the better educated (see Table 5.4).

Age shows less stratification in VCR ownership than either social class or education, which are related. However, in Barquisimeto, there was a tendency for VCR ownership to be concentrated most among people aged 30 to 50 (see Table 5.5).

In Venezuela, as elsewhere, social class, viewing interests, and the status of popular culture on television are related. As the above survey figures indicate, few lower-middle or lower-class households have

TABLE 5.4   VCR OWNERSHIP AND EDUCATION IN BARQUISIMETO, VENEZ., 1985

	Illiterate some primary	Primary complete	Some secondary	Secondary/ some univ.	complete univ.
Own VCR	3 %	10 %	10 %	14 %	43 %
No VCR	97	90	90	86	57
Total	(176)	(148)	(146)	(88)	(47)

Source: From Some remarks on the agenda for communication research in the 80's: The impact of the new information technologies on Latin American societies by F. Izcaray, 1984, paper presented at the International Association for Mass Communication Research meeting, Paris.

TABLE 5.5   VCR OWNERSHIP AND AGE IN BARQUISIMETO, VENEZ., 1985

	Age 0–30	31–40	41–50	51–60	61–over
Own VCR	5 %	14 %	15 %	10 %	8 %
No VCR	95	86	85	90	92
Total	(149)	(169)	(138)	(96)	(53)

Source: From Some remarks on the agenda for communication research in the 80's: The impact of the new information technologies on Latin American societies by F. Izcaray, 1984, paper presented at the International Association for Mass Communication Research meeting, Paris.

VCRs. Unlike some countries, particularly those in Asia and those of the Arab world, group viewing in extended families, village centers, public places, and commercial establishments does not dramatically extend the reach of VCRs in Venezuela, according to in-depth interviews by the author with a number of VCR users in 1985.

Interviews with viewers, industry and academic observers, and videotape rental shop operators in Caracas and Barquisimeto also revealed major divisions in taste between upper-, middle-, and lower-class audiences. In general, upper- and upper-middle-class Venezuelans may have acquired VCRs precisely because they were less satisfied with the current content of broadcast television. Their taste and education seem to be more internationalized, to the extent that they more often prefer to watch imported films and television series, even in English, than Venezuelan Spanish-language television programming, which is shifting away from the interests of the upper and middle classes toward a more mass audience. Research shows that the percentage of television time in Venezuela devoted to imported programs has decreased from 50 percent in 1972 to 33 percent in 1982 (Antola & Rogers, 1984). To the degree that some upper-class

Venezuelan audiences want to continue watching imported television programs and films, VCRs provide an option.

In contrast, both audience ratings and in-depth interviews show that lower-middle and lower-class Venezuelan audiences are more content to watch what is currently broadcast. In prime time, that consists primarily of *telenovelas*, comedies, and musical variety shows, with some imported series and films dubbed into Spanish. The increased proportion of the Venezuelan television audience represented by middle- and lower-class groups seems to parallel the increased proportion of the content of television taken up by Venezuelan productions and Spanish-language "regional" material imported from other Latin American countries (Izcaray, 1984; interviews, J. M. Aguirre and Marcelino Bisbal, professors, Universidad Central de Venezuela, Caracas, Venezuela, June 18, 1985).

Overall, the case of Venezuela indicates several uses of VCRs for entertainment. All viewers can use VCRs to more conveniently watch imported feature films. Beyond the usual consumption of feature films, internationalized television viewers, primarily in the upper or upper-middle classes, can reverse the trend in broadcast television toward more Latin American programming.

Colombia

As in Venezuela, possession of a videocassette recorder in Colombia became a major middle-class status symbol in the late 1970s and 1980s ("Guerra al Betamax," 1984). Also, as in Venezuela, VCRs had an initial surge of popularity, because they offered a way to see color television before it was introduced as a broadcast system in the early 1980s. In both Colombia and Venezuela, competition from color videocassettes seems to have pushed the governments into allowing the introduction of color broadcast television (Brito, 1985).

Also as in Venezuela, there were no tariff, tax, or import restrictions imposed by government to raise prices on VCRs or reduce their availability. In both Colombia and Venezuela, this made VCRs more widely available to a broader range of potential purchasers than in many other Latin American countries (*NTC/NCT Newsletter*, 1986, pp. 11–12).

There are about 2 videocassette recorders for every 100 Colombians, covering 17.7 percent of television households. Press reports from Colombia indicate that, as in Venezuela, VCRs are concentrated in the upper and upper-middle classes but are a primary status symbol and consumer goal among the entire middle class (interview, Azriel Bibliowicz, professor, Universidad Javeriana, Bogotá, in Minneapolis, MN, October 1984; "Guerra al Betamax," 1984; Stangelar, 1985).

More than in most of the other Latin American countries considered here, local observers in Colombia, such as Bibliowicz (1982),
attributed the popularity of VCRs to a lack of diversity in broadcast
television. Depending on the region, there are 1 to 3 channels available, all owned by the government. The content is intended to emphasize education, and does so to a considerably greater degree than other
Latin American countries, according to statistics reported to UNESCO
(1985) and analyzed in Straubhaar (1985). At least two of the channels
do emphasize entertainment in evening prime time, under an arrangement where time is rented to private commercial program services (Fox
de Cardona, 1975). Nevertheless, diversity of choice is perceived to be
much lower on broadcast television than on radio, for example, leading
those who can afford it to seek viewing options through VCRs
(Bibliowicz, 1982).

Brazil

In Brazil, the situation for VCR adoption and use is very volatile,
having changed a great deal in 1986. VCR penetration is still relatively
low. Compared to either Colombia or Venezuela, there are fewer VCRs
in Brazil as a proportion of the population (1.5 percent) or television
households (6.3 percent) (see Table 5.2). Still, in 1985, the estimated
number of VCRs was only about 2.3 percent of television households,
showing dramatic growth in 1985–1986.

Until 1986, Brazilian television industry sources indicated that VCR
ownership was almost entirely contained within the upper and upper-
middle classes: 88 percent in Classes A and B, 12 percent in C, D, or
E (IBOPE, 1984) in the cities of Rio de Janeiro and São Paulo. Still
manufacturers anticpated that the market would broaden considerably
in the next few years (*Veja*, 1981; Stangelar, 1985; interview, Roberto
Marcelo, director, Globovideo, Rio de Janeiro, Brazil, July 16, 1986).

It is useful to try to explain why VCR diffusion in Brazil prior to
1986 was low, despite the fact that Brazil had a higher average income
than Colombia, where VCR penetration was, and is, much higher (see
Table 5.2). The first reason is that Brazil, as noted above, prohibited
the importation of VCRs (which raised prices despite smuggling) and
heavily taxed nationally produced VCRs. The result was VCR prices
generally twice that of the world market. The second reason seems to
be that more diversity exists within what is broadcast on television in
Brazil than in some other Latin American countries, so that fewer
potential viewers felt the need to acquire VCRs as a substitute for
television. In particular, Brazilian television has been segmented by
social class, which tends to broaden viewing options and reduce the
need for VCRs, even among smaller parts of the audience.

In most Brazilian cities, audiences can choose between four networks that target different audience groups. TV Globo's general audience programming is a broad representation of current Brazilian television popular culture, with *telenovelas*, Brazilian-produced action and dramatic series, musical variety shows, comedy shows, and news programs. TV Bandeirantes and TV Manchete target the upper-middle and upper classes, with imported series and feature films, and more news, public affairs, and "serious" entertainment (highbrow miniseries and issue-oriented music and comedy shows). SBT, another Brazilian network, programs for lower-class tastes with game shows, variety shows (*shows de auditório*), talk shows, sentimental *telenovelas* (often imported from Mexico), and differently targeted music and comedy shows (Litewski, 1985). In this constellation of diverse offerings, even the elite's demand for international feature films and series is somewhat satisfied, so it seems reasonable that relatively few have gone to the extra expense to acquire VCRs to get more film material.

The factor of cost has also been significant in keeping down VCR penetration in Brazil. The Brazilian government has embargoed imported VCRs to stimulate production of VCRs in the Manaus Free Zone. However, VCRs produced there cost U.S.$800–$1,500, while a VCR smuggled in from New York or Panama may cost as little as U.S.$400. Nevertheless, access to the smuggling options is limited to those who travel or know how the smuggling operations work—that is, the upper and middle classes (Mattelart & Schmucler, 1985). The economic picture for VCR acquisition changed considerably in 1986. The wage-price freeze inaugurated in March 1986 by President Sarney significantly increased purchasing power, at least among the middle class, leading in part to a boom in VCR purchases. Even the more expensive nationally produced machines were sold out by late 1986 ("Videocassete no Brasil," 1986).

Most VCR use in Brazil is watching prerecorded movies. In a 1983 survey of videoclub clients, 88 percent said they used VCRs to watch movies, 64 percent to record TV programs off the air, and 24 percent to make their own home recordings (Santoro, 1985). Still, it seems that a number of alternative, nonfeature-film-oriented VCR uses are also driving the recent explosion of VCRs (from an estimated 500,000 in 1985 to over 1 million—probably 1.2 to 1.3 million—in 1986). Much of this use seems to be home recording, extended family and group recording, and the purchase/rental of how-to-do-it and information-oriented tapes. Although somewhat unclear because of smuggling, the ratio of cameras to VCRs among private owners seems higher than in the United States. Roughly 20 percent of VCR owners have video cameras (Santoro, 1985). There are also specialized production companies that tape births, sports events, or weddings. Sports and other how-to tapes, plus music- and arts-oriented videos are in commercial circulation. All of these types of programming—almost all produced in Brazil—are currently in rapid expansion. Some applications and ventures will fail, but the trend seems away from the use of VCRs only for time shifting and feature film playback ("Videocassette no Brasil," 1986; interview, Rosalee Bloch, director, Manchete Video, Rio de Janeiro, Brazil, July 16 1986).

There are also politically oriented alternative uses of VCRs in Brazil. A variety of groups that feel excluded from access to other media have turned to VCRs as an alternative channel, both for internal communication and for spreading their message to others. These groups include the Roman Catholic Church, labor unions, community and neighborhood associations, cultural groups, and others (Festa & Santoro, 1987).

The Dominican Republic

The Dominican Republic has an extremely low VCR penetration rate; 2.5 percent of television households, 0.2 percent of the population. The rate is somewhat higher in the capital of Santo Domingo, where 4.4 percent of television households have VCRs (Straubhaar & Viscasillas, 1987). This seems to be related to both the ability to purchase a VCR and the availability of other video media to supply diverse types of programming.

The poverty of the Dominican economy, particularly notable during recent years when VCRs have spread elsewhere, constrains the ability of all but a few Dominicans to buy them. Per capita income is lower, and income distribution more skewed to concentration in few hands, than in the other countries considered here (see Table 5.2; World Bank, 1985).

Even among those Dominicans who can afford a VCR, relatively few seem to have acquired one. Interviews by the author among journalists, media industry people, and government officials found only a few who had VCRs. Of forty-five journalists interviewed in 1984, only three had VCRs: two used them only for professional purposes (recording competing news shows to watch later), and the third found that he did not use his for entertainment, although that was why he had purchased it. Similarly, in 1987, three of twelve people (in a survey of 250 television households) who had VCRs, said they did not use them (Straubhaar & Viscasillas, 1987).

A somewhat larger number, about ten out of forty-five jounalists interviewed in 1984, and 25 percent of TV households surveyed in

1987, subscribed to one of the two English-language cable systems available in upper-income neighborhoods. (From 1984 to 1987, cable reached approximately 2 percent of the households in Santo Domingo.) Cable TV in Santo Domingo supplies HBO (Home Box Office), Cinemax, MTV (Music Television), ESPN (Sports), CNN (Cable News Network), Disney (cartoons, films), and over a dozen other channels pirated from United States distribution satellites. For those with access to cable TV, this provides considerable diversity of programming along specialized interest lines. This limited access to cable television again reflects the increasing segmentation of the audience along class lines (interviews, Jorge Blanco, independent producer; José Cabrera, Secretario de Comunicación Social; and participants in journalism seminar, Universidad Católica Madre y Maestra, Santo Domingo, Dominican Republic, December 1984).

As in Brazil, a large number of broadcast channels are also available in most Dominican cities. Santo Domingo has six channels in Spanish: five private, one government-operated. Most of those interviewed thought that sufficient diversity was offered through these broadcast channels that VCRs and even low-cost pirated English-language cable programs would have relatively little appeal to any but a very restricted internationalized economic and educational elite. This is confirmed by a survey of TV households in 1987, which showed that actual use of English-language channels tended to be restricted to those that required relatively little English-language ability, such as MTV, ESPN, and Disney. In general, upper-class audiences watched English-language cable more, but actually spent most of their viewing time on Spanish-language material that was either produced locally or imported from other Latin American countries (Straubhaar & Viscasillas, 1987).

VCR REGULATION

In Argentina, Brazil, Colombia, Mexico, Panama, and Venezuela, a new set of industries is being created to supply VCRs and video cassettes. To date, these new industries have competed directly with existing television broadcast and film distribution industries. They also challenge the economic and political goals of several governments.

The new video industries have largely been illegal or "informal." Both VCRs and videotapes are frequently imported and distributed illegally. In Brazil, for example, over 90 percent of prerecorded tapes, 80 percent of VCRs, and 60 percent of blank videotapes are imported illegally (Santoro, 1985). At first, this was mostly because legal sales and distribution in these countries were prohibited or restricted, except for countries such as Colombia, the Dominican Republic, or Panama. Brazil made importation of both VCRs and videotapes illegal and enforced the prohibition, because the government perceived them as unnecessary luxury consumption that contributed to balance of payments problems. Colombia placed restrictions on VCR imports but did not enforce them vigorously, while Venezuela did not restrict VCR purchases. Neither Colombia nor Venezuela legalized prerecorded tape importation initially, apparently in hopes of restricting the spread of the machines (interview Azriel Bibliowicz, professor at Universidad Javeriana, Bogotá, Colombia, in Minneapolis, MN, October 1984; "'Legal' homevid," 1985). The Dominican Republic and several other smaller countries have created no regulations or laws concerning VCRs or videotapes.

These regulatory approaches have had differing results. Brazil was somewhat successful in restricting the importation of VCRs while economic recession kept demand low, but much less successful when the consumer economy heated up in 1986. None of the countries has been successful in restricting the illegal entry of software or prerecorded videotapes (most of the latter are illegal or pirated copies of North American feature films; see Table 5.1). Furthermore, the MPEAA and national film producers complain that the lack of regulation in most Latin American countries has contributed to the problem of piracy and hurt their revenues.

VIDEOCASSETTE PROGRAMMING

VCRs tend to break "the monopoly on the conversation" that broadcast television has traditionally exercised (Sodré, 1977). While VCR users may record off the air, most in Latin America choose instead to use VCRs to substitute prerecorded video material for that which is broadcast. General estimates cited by Mattelart and Schmucler (1985) showed that VCR users in Latin America tended toward playback/ substitution (76 percent) more than those in the United States (48 percent) or Western Europe (41 percent).

If Latin Americans largely use their VCRs to substitute new material for what is broadcast, as the author's interviews in Brazil, the Dominican Republic, and Venezuela confirm, then what do they choose? A concise answer given by one videotape rental shop owner in Barquisimeto, Venezuela, was, "violence, pornography, and kids' shows...from the United States." O'Sullivan (1985) said "specialists estimate that close to 40 percent of the videocassette industry in Latin America today is pornographic; and while these cassettes mostly come from United States sources, they also include a considerable part of the European production." One should note, however, that other sources, such as Santoro (1985), estimate the proportion of pornography among videocassettes in circulation to be much lower.

A variety of sources in most of the countries studied indicate that most VCR rental fare is imported feature films from the United States. In Brazil, 80 percent of rental videotapes are from the United States, according to a survey of rental catalogs by Santoro (1985). Brazilian, Japanese, and European films are also available. In Brazil, the primary kinds of video programming were entertainment, including adventure, drama, and comedy (67 percent); children's (10 percent); music (20 percent); and pornography or erotica (2 to 5 percent). In Venezuela, beside United States movies, Venezuelan films were also available but almost none from other Latin American, European, or Asian countries. In Colombia, some limited variety of Latin American and European films existed. In Panama, a wide variety of both United States and Latin American films was stocked, partially for re-export.

Data on actual viewer choices for videotape rentals are more limited, but the direction can be seen by what rental shops and clubs choose to stock in response to client or member demand. A 1987 survey in the Dominican Republic showed that, of a limited VCR-owning population (4.4 percent of TV households in Santo Domingo), eight out of twelve VCR owners used them mostly for imported films, one for musicals, and three did not use the machines much (Straubhaar & Viscasillas, 1987).

All of this information about prerecorded tape choices, with the exception of nationally produced films in Brazil and Venezuela, creates a fear that VCRs will become primarily a channel for imported films. In this event, VCRs would be a new channel for cultural dependency, given the overwhelming dominance in the present market of films from the United States. This new channel arises at the same time that television broadcasters in all these countries are increasingly showing nationally produced programs or programs imported from other Latin American countries (Antola & Rogers, 1984; Straubhaar, 1984).

VCRS AND ALTERNATIVE COMMUNICATION

While most individual VCR owners in Latin America still use their machines for playing prerecorded entertainment tapes, there seems to be a growing use of VCRs to produce alternative communication. The largest part of this kind of video production and distribution is among popular groups.

> Popular groups as understood in the Latin American context refer to representatives from the poorer sectors of society . . . . For the popular movements, video is an instrument of the expression of their culture, an alternative culture to the one represented on the mass media. It represents the search for an artistic language that expresses the cultural values of the people. It is also a medium that contributes to the political education of the popular movements (O'Sullivan-Ryan, 1985, p. 25).

In another formulation, surveying a variety of productions assembled by Democracy in Communication, Karen Ranucci (interview, Philadelphia, PA, October 1986) sees three functions for popular video. First is empowerment; both urban and rural people gain confidence in the initial learning of video production skills. Second is the sharing of experiences with other groups; for example, sharing tapes on alcoholism, reading, health, family counseling, and skills such as basket weaving. Third is creating a broader network that may eventually democratize and decentralize communication.

The degree of political openness in Latin American countries is also crucial in fostering alternative video. According to a report by IPAL (Center for Study of Transnational Culture):

> Parallel to the commercial and domestic use fo video, efforts have been made to use video for individual and collective expression, particularly in contexts where political repression has closed the mass media to democratic opposition and to grass-roots movements. The circulation of information is limited, as there is little interaction between the groups involved. The degree of political repression on the one hand, and the degree of organization of the grassroots movement on the other, have direct impact on this (video) development (Festa & Santoro, 1987).

One of the institutions most active across Latin America in promoting the alternative production and use of "popular" video is the Roman Catholic Church, particularly the Latin American section of the International Catholic Organization for Film, the UNDA, and UCIP. The Latin American Bishop's Council (CELAM) has organized a center in Bogotá to promote, but not actually do, the production of video. While church groups are using VCRs for formal education and evangelization, the most common use, according to a survey of Latin

American Catholic groups by O'Sullivan, is for "popular promotion. . . by Christians in alliance with movements among the poor and marginal groups" (1985, p. 25). He gives an example of an Argentine priest who uses a VCR to help farmers in a dispute over land titles: taping meetings, developing strategies, honing arguments, and integrating the issues into local school evangelization programs (p. 25). Similarly, Festa and Santoro (1987) think that the church's use of video, starting with its urgings at the Third Puebla Conference, is mostly for evangelization, but that has set the stage for other church-group uses at the grass-roots level.

Another kind of VCR use is for training adults and children to become more active and critical viewers. O'Sullivan reports the use of video playback and discussion, and the actual hands-on use of equipment, to help demystify the media and to understand media language. In Brazil, Regina Festa, training people from the Metal-workers Union to produce videos, reports that a major product is the feeling of understanding and empowerment vis-à-vis television that people get, particularly after they learn to edit. She reports one worker commenting, "Hey, we are just as good as TV Globo [the major commercial network]" (Regina Festa, professor of journalism, Universidade de São Paulo, Brazil, interview with author, July, 1986).

In both Nicaragua and El Salvador, VCRs play a role in revolutions. In El Salvador, both the revolutionary and government armies use portable VCR screenings for propaganda. In Nicaragua, the Sandinista government and groups friendly to it use video to consolidate the revolutionary government and to keep people loyal, in the face of efforts by the *contras* to create a counterrevolution.

Alternative Video in Nicaragua

The Sandinista government has embraced video production, both for broadcast and VCR distribution/showing, as a primary means of communication. The use of small-systems video, or portable VCRs and cameras, has enabled a number of groups to begin recording a wide range of community events, producing documentaries and even "news" coverage. The activity comes from a series of production groups, including the Sandinista-controlled television channels, government ministries, affiliated political groups, and independent or private community, church, or work-related groups.

For example, the Ministry of Agrarian Reform (Midinra) has a video unit that concentrates on the effects of government agricultural and economic policies. One of its productions was a seventeen-minute investigative report on shortages entitled, "What's Happening to the

Toilet Paper" ("Que pasa con el papel higienico?") ("Democracy in Communication," 1986). As one of the best-equipped organizations, Midinra also is involved in other projects such as the production of "Sandino Vive," a tape for the fiftieth anniversary of Sandino's death. Midinra pays the official television channel Sistema Sandinista, to air monthly programs, some of which have been rejected (Halleck, 1984, p. 15), reflecting a degree of decentralization common in video production in Nicaragua. Midinra supplies most of the videos that are shown outside Nicaragua, through groups such as X-Change TV. Exchange activities with United States groups have also helped stimulate production in Nicaragua, particularly in the early days of the revolution, 1979–1981, when a number of United States and European video makers helped teach production skills.

Other production groups include the Taller Popular de Video Timoteo Velásquez, linked to two major unions, the Central Sandinista de Trabajadores and the Asociación de Trabajadores de Campo. More official groups include Incine, the film/video production unit of the Ministry of Culture, and the audiovisual department of the Ministry of the Interior, which produces broadcast versions of the weekly public meetings between the *comandantes*, the nine leaders of the government, and urban and rural groups. Those in Managua are broadcast live, with remote equipment (Halleck, 1984).

These meetings, called "Cara al Pueblo," are considered to be crucial for communication with government leaders. Depending on the account, some people consider them to be open and effective upward communication, while others feel they are stage-managed by the local committees for the Defense of the Revolution, with only acceptable questions permitted (Halleck, 1984). One of the issues in debate currently is how much questioning of government policies should be permitted. A more culturally oriented program that permits local expression is the traveling game show, "Aquí en esta esquina" ("Here on This Corner"), produced by the Sistema Sandinista de Televisión, live from different neighborhoods.

In both cases, the government is using simple portable equipment to create more local and, it is hoped, more responsive, programming. As Halleck (1984) notes, before the revolution in 1979, only news programs were produced in Nicaragua; all else was imported. VCRs and portable broadcast equipment have at least changed that, since now about four hours are produced daily (interview, Martha Wallner, Democracy in Communication, Philadelphia, PA, October 1986). Given the diversity of actors and producers, which the technology facilitates and the political system so far permits, it seems that VCRs and video in Nicaragua are opening and decentralizing television.

Revolutionary Video in El Salvador

Probably the main use of VCRs in El Salvador is still watching prerecorded tapes. But the country provides an interesting example of VCR use in war propaganda. The Frente Marabundo de Liberación Nacional (FMLN), the main antigovernment guerrilla group, has had a video collective for six years. The FMLN produces and distributes tapes both inside and outside its own zone of control, which includes the capital of El Salvador.

Some videotapes are produced only for the use of the FMLN's own cadres and forces and include meetings, speeches by commanders, and the like. This permits an effective extension of the reach of meetings. The FMLN also produces video documentaries and propaganda to reach supporters and undecided people outside its own ranks (interview, Carlos Guery, video producer from El Salvador, Philadelphia, PA, October 1986). FMLN tapes are particularly aimed at rural people in areas outside the war zones but are also intended to help win support in the United States and elsewhere outside El Salvador ("Democracy in Communication," 1986).

Alternative Video in Brazil

Although groups are active in Bolivia, Chile, Mexico, Peru, Panama, and Uruguay, Brazil probably has the largest number of independent video producers in Latin America. Experimentation dates back to the introduction of half-inch portapacks in the early 1970s.

Santoro (1985) and Festa and Santoro (1987) feel that the relaxation of television censorship after the opposition party victories in major states and cities in 1982 contributed to more video production. They also feel that as new producers are trained and discover the limits in the broadcast television marketplace, many move into video production as a professional alternative. But most of all, the major media still fail to address many of the major issues or cover popular movements; thus these movements seek "vindication and debate" through video that reaches an organized interested public and sometimes even gets on the air. The publication *Video Popular* (circulation 2,000) is one effort to increase awareness of what is being produced by other groups.

The fragmentation of groups producing alternative video has led to a series of meetings from 1982 on, and the eventual organization in 1985 of the Brazilian Association of Grassroots Video Movements. This association promotes debate about the democratization of communication, publicizes video productions, organizes sales and distribution of videos, and helps train and supply equipment to groups. About 100 individuals or groups involved in the association have produced about 400 video projects, plus a large quantity of recorded, but unedited, material (Festa & Santoro, 1987).

In Brazil, the exchange and circulation of independently produced videos is a major problem. The Brazilian Association of Grassroots Video Movements has its "*Catalogue of Grassroots Video Programmes,*" listing over 100 productions. Exchanges are also being sponsored by the Cultural Department of the current government of the state of São Paulo, which subsidizes the Brazilian Association of Grassroots Video Movements and has proposed putting at least one VCR in all public libraries (Santoro, 1985).

Video seems to be used in six principal ways: (1) self-oriented, for internal use, with no outside distribution planned; (2) simple recordings to preserve events for later reference; (3) rough documentaries based on original material—usually group projects; (4) planned documentaries; (5) original scripts with a planned narrative structure, including some fiction; (6) prerecorded tapes from other popular groups (Festa & Santoro, 1987).

The types of production in Brazil are varied, including carefully scripted dramas of 40 to 60 minutes, music videos, collages of image and music, festival and carnival footage, caricatures of documentaries and TV newscasters, speeches by union and political leaders, alternative documentaries, and feminist productions, among others. Some tapes are quite widely and successfully distributed. For example, the Metalworkers' Union produced a video, "*Abrindo o Pacotão*" ("Opening the [Economic] Package"), which was copied and distributed in fifteen Brazilian states and seen by thousands of union leaders and members (Festa & Santoro, 1987). In fact, the regular productions of Workers' TV, a series of productions by the metalworkers' unions of São Bernardo and Diadema, near São Paulo, have probably been the most significant, in terms of both political effect and systematic exploration of alternative narrative structures, systematic training of workers in production by professionals (Festa & Santoro, 1987; interviews with Regina and Festa Luis Fernanda Santoro, Professors of journalism, Universidade de São Paulo, São Paulo, Brazil, July, 1986).

VIDEOCASSETTE TECHNOLOGY

In Brazil, 90 percent of the VCRs and videotapes sold are in the VHS standard, 9 percent U-Matic, and 1 percent Betamax (Santoro, 1985). VHS is also the dominant VCR format in Argentina, Chile, Paraguay, and Uruguay, while Betamax is dominant in Colombia, Peru, and Venezuela (*NTC/NCT Newsletter*, 1986, p. 5).

Another technical factor complicating video software distribution is the split between the NTSC and PAL line/color systems for television in Latin America. Argentina, Paraguay, and Uruguay have PAL; Brazil, PAL-M; while most of the other countries, including Bolivia, Chile, Colombia, Ecuador, Mexico, Panama, Peru, and Venezuela, have NTSC. In the countries that do not have NTSC, the direct importation of videotapes from the United States is made complicated by the difference in standards. In Brazil, the typical response has been to either build dual-standard machines that use both PAL-M and NTSC, or modify either PAL-M or NTSC machines to play the other standard as well. The modification costs U.S.$70 to $80 and has become a standard part of the price for importing or smuggling in VCRs (Santoro, 1985).

VIDEO MEDIA COMPETITION: VCRS, DBS (DIRECT BROADCAST SATELLITES), CABLE, AND TV

In Western Europe, Japan, Canada, and the United States, the VCR must be analyzed in terms of competition and complementarity with other video media. This is obviously true of broadcast television and VCRs in Latin America. VCRs are too expensive, relative to average income, for most prospective buyers in Latin America to purchase them primarily for time shifting. Furthermore, the quality of broadcast television is frequently high, particularly in providing acceptable entertainment. Thus, the "marginal utility" of a VCR to a poor consumer, in providing basic entertainment and information, may be lower than in regions where entertainment is not as prominent.

Among "new" video technologies, beyond those of broadcasting, VCRs are preeminent. Cable TV, for instance, is becoming prominent only in the Caribbean and Central America, although it is beginning to spread in Buenos Aires and Mexico City and is under planning consideration in Brazil, Colombia, and Venezuela. Home satellite dishes are spreading in Central America and the Caribbean. Both cable TV and satellite dishes in areas relatively near the United States are driven by the possibility of pirating programming directly from the United States distribution satellites that feed cable TV. The move toward scrambling such signals may change the situation, as may efforts by the MPEAA, the cable program services, the satellite companies, and the United States government, to force such pirate operations to legalize and pay copyright royalties.

THE IMPACT OF VIDEO ON TELEVISION INDUSTRIES

One of the impacts of video in Brazil and Nicaragua has been to open up the possibilities for broadcast television production. The use of VCRs and portable remote equipment by Nicaraguan broadcasters has been discussed above, but it is worth repeating that some observers attribute much of the increase in national production of television under the Sandinistas to video. As in Nicaragua, several small Brazilian production companies are building on experience in independent or alternative video production, using lightweight portable equipment to produce programs for television networks as well. The Olhar Electrô-nico (Electric Look) group produces programs for TV Gazeta in São Paulo, and Goulart de Andrade did the youth-oriented program, "Perdidos na noite" ("Lost in the Night") for TV Record (Santoro, 1985; "Democracy in Communication," 1986). Some community groups, such as the metalworkers' unions of the São Paulo region in Brazil, also hope to build on VCR production to appeal for broadcast television licenses, if the process of license applications is opened up to community groups. This was, in fact, an issue under consideration in 1986–1987 by the Brazilian constitutional convention.

Another implication of VCRs for broadcast television is more institutional in nature. Since most of the videotapes watched in Latin America are pirated copies of foreign films, this has created a conflict between United States film distributors, local broadcasters, and video rental shops and viewers. Some proposed solutions strengthen the position of American film distributors and of the national media oligopolies, which operate in television, film distribution, and now videotape distribution.

However, television industries also fear some economic effects. When VCR users substitute rented films for television viewing, the television audience decreases and advertisers pay broadcasters less for the smaller audience that is delivered to them. This is particularly acute if VCR use causes precisely the most sought-after advertising audiences, the upper middle and middle classes, to watch less broadcast television. It seems that in some countries, particularly Venezuela, the VCR may have the effect of differentiating (or further differentiating) the broadcast audience into class-based segments. In the extreme case, broadcast television might increasingly become a middle- and lower-class phenomenon, while the upper classes use VCRs to watch material of their own choice. In Brazil and Venezuela, television networks are reacting by getting into videotape distribution, resulting in a further integration of cultural industries.

Even if commercial television networks cope with the VCR threat by getting into video distribution, as both TV Globo and TV Manchete in Brazil are doing, there is still a question of fundamental change in the traditional impact of broadcasting systems. One of the worst fears, particularly for those who had hoped to see broadcasting become a public service medium in Latin America, is precisely the fragmentation of the mass audience—changing a shared national experience into a set of "narrow-cast" exchanges with little common or shared culture.

THE IMPACT OF PIRACY ON CINEMA PRODUCTION AND DISTRIBUTION

Partially because most of the videotapes circulating in Latin America are, in fact, illegal copies brought in from outside, the main economic challenge posed by VCRs is to existing feature-film distribution channels, both national and international. The Motion Picture Export Association of America (MPEAA) has been the main actor because it represents the North American film studios, which produce most of the material that has been pirated. The MPEAA has estimated that North American film producers have lost U.S.$700 million in potential gate receipts because potential viewers have watched films via pirated videotape copies instead. In Venezuela, for example, such losses by MPEAA distributors to video piracy were estimated at U.S.$20 million just for 1981 ("'Legal' Homevid," 1985).

The MPEAA has been putting considerable pressure on Latin American governments to legalize the importation of legitimate copies of feature films, and this has brought about some changes in policy. Venezuela legalized and regulated the importation of films on tape in 1983. The Brazilian government began negotiating with the MPEAA and other film distributors in 1984–1985 to permit the legal importation of films on tape, under fairly strict controls. The rules require the inclusion of Brazilian short films on cassettes to stimulate Brazilian production. In 1985 the Colombian government was also moving toward legalization and regulation ("'Legal' Homevid," 1985).

But the creation of legitimate channels for the importation and the rental or sale of videotapes and VCRs has not ended illegal imports.

Pirated tapes are usually much cheaper to rent or purchase in all these countries. In Venezuela, the purchase or the exchange or rental of pirated tapes was a quarter of the cost of legitimate copies (interviews with rental shop owners, 1985). Pirated video copies of films have usually been available much sooner than legal copies. Particularly in the past, illegal copies have been available months or even years before films were available in theatrical distribution, and even longer before legal video copies were available. In some cases, notably Brazil before the recent change back to civilian rule, some kinds of films were not available at all, except in pirated video copies. This has been less of a motivation in most Latin American countries than in other Third World countries, where censorship is much more extensive.

The problem of piracy is also acute for Latin American national film industries. As documented in India and other countries, VCR use seems connected to a falling-off in cinema attendance, which further hurts the financial basis of the film industry. This connection has not become a major point in Latin America yet, perhaps because film producers have been struggling with more serious coincidental problems. The costs of producing films in Latin America have steadily increased. National producers are in essence guaranteed distribution through systems that require national films to be shown a certain number of days in theaters in most countries. Nevertheless, the number of films produced has declined. In some countries, such as Brazil, a further problem has been the decline in production of general audience films and a shift toward the production of pornography (Johnson, 1986).

Still, in the countries that have film production industries, some concern about video piracy centers on feared damage to the domestic industry. Several countries, such as Argentina and Brazil, have concentrated their legal efforts on copyright protection for national film makers. As a result, far more legal copies of national films than of foreign films are available in both Argentina and Brazil. Legal distribution companies have grown up in large part to handle national films. In Argentina, those tend to be independent; but in Brazil, as in Venezuela, video distribution of feature films is being taken over by video distribution branches of broadcast companies. In Brazil, the two most active are Globovideo, a branch of TV Globo, and TV Manchete.

However, since most countries import most of their films, the damage primarily accrues to cinema owners, distributors, and foreign producers and distributors. In several countries, efforts by foreign and local distributors and producers to cut piracy are reshaping film industries, or at least film distribution (see Appendix 5.B).

LEGAL VIDEOTAPE DISTRIBUTION
AND OLIGOPOLY IN VENEZUELA

Efforts are underway to create legal video distribution channels in Brazil and Venezuela. Because of its larger population of VCRs, the MPEAA pressured Venezuela first to crack down on illegal video distribution. An initial judicial decision created a precedent for seizing stocks of illegal copies and arresting those involved in illegal video distribution sales or rentals, although continuing complicated litigation may reverse the antipiracy law enforcement ("MPEAA Hits Caracas Pirates on Copyright," 1983). Police raids have reduced the willingness of rental/purchase shops to use pirated material, since the seizure of an estimated 30 million illegal copies has been costly for distributors (Scharfenberg, 1985).

Furthermore, MPEAA members have advanced their schedules for releasing films on video in Venezuela, to ensure that legal copies are available on the market in time to compete with illegal ones. Finally, the MPEAA has orchestrated a publicity campaign about the superior viewing quality of legal copies over pirated copies, which are, in many cases, fourth- or fifth-generation reproductions.

Along with the crackdown on pirated copies, the MPEAA, Venezuelan film distributors, and major television networks are working together to create legal distribution mechanisms to control the video rental/sales market. CIC Video (Paramount and Universal Studios) has arranged with theatrical distributor Tony Blanco to create the video distribution chain, Blancivideo, which now has a major share of the rental sales market. Warner Brothers and the Cisneros family, owners of Venevision (one of two major commercial networks), have created Video Rodven. Other distributors are also forming multinational ties, such as that of Video Express to VCL Video of Great Britain. At least one smaller distributor, Bustillos, fears that a narrow oligopoly structure for both theatrical and video distribution of films will be formed between North American film sources, Venezuelan television networks, and Venezuelan film distributors (Brito, 1985; interviews, Marcelino Bisbal and J. M. Aguirre, Caracas, Venezuela, June 18, 1985).

VCRS AND THE AUDIENCE

There is a great deal that is not yet known about the effects of VCRs on the audience in Latin America. Most of the studies to date have been structural and institutional, such as those reported in *NTC/NCT Newsletter* (1986).

Among those in Latin America who have commented on the new medium, the anticipated effects of VCRs on audiences and on popular culture have created both considerable hopes and fears (Stangelar, 1984). The main mechanism underlying both hopes and fears is the personal control individual viewers have over the video medium. The problems center on what options individuals will have to choose from and what choices will be made.

The rise of VCRs as a new channel for cultural imports, at the same moment that the content of broadcast television is being increasingly nationalized or regionalized, may be no accident. At least one partial connection seems to lie among viewers' social class and related interests. At least in Brazil and Venezuela, it is precisely the upper- and upper-middle-class viewers, who are dissatisfied with the recent changes in the content of broadcast television (interviews, Clemente Cohen, Caracas, Venezuela, June 1985; and A. Machado, marketing director, Instituto Brasileiro para Estudios de Opinião Publica e Estatistica, Rio de Janeiro, Brazil, November, 1982).

The programs in which the nationalization or regionalization of television content has taken place in Latin American countries are *telenovelas*, local musical variety shows, game shows, and comedy. These have been particularly popular with middle- and lower-class audiences, who now dominate the demographics of television and have, consequently, changed the advertising and programming strategies of commercial broadcasters (Sodré, 1977). Upper-class viewers have shown a much more marked preference for imported action adventures, situation comedies, and feature films (Straubhaar, 1981). Thus, VCRs may reflect and facilitate a division in what had been thought of as the mass audience for television in these countries. However, as the current Brazilian experience shows, that segmentation can take place even within broadcast television, limiting the role for VCRs.

The segmentation of the audience, whether by diversified broadcasting, VCRs, or other new technologies, such as cable TV or direct broadcast satellites, is threatening in some aspects. Antonio Olivieri (1985) asks, "Will it be good or bad if we can choose freely between the television of Fidel Castro, Billy Graham, and Col. Khadafy . . . and *Playboy*?" Cultures may fragment under such a choice. In particular, cultural production in some small- and medium-sized countries may suffer if the audience, and with it, the resource base for production, is split in too many ways.

Nevertheless, if the average television viewer has only an increased quantity of imported films to choose from, then it seems that many viewers, particularly in the middle and lower classes, are likely to stay with broadcast television. One should not underestimate the continued attraction to the audience of the version of popular culture that has been evoloving on television in the major Latin American countries. The data cited above on the relatively limited popularity of feature films and other imported programs on both VCR and cable TV in Santo Domingo indicate that new technologies will not necessarily change existing television programming and viewer preference patterns, at least not for the majority of the mass audience. The elites and the upper-middle class, as our data indicate, may well use new technoligies to choose imported entertainment to a much greater degree.

If institutions such as the Roman Catholic Church lend their resources and organizational networks for exchanging community productions, then "popular" programming from a variety of points of view, about a variety of issues, may also have a chance to gain some part of the new video audience. The audience may even have the opportunity, long awaited by critics, to talk back. However, because audiences have been conditioned for years to expect certain kinds of television, it may be hard for community video producers to regain much of an audience; but a few innovative video and television experiences in Brazil, Chile, and Nicaragua indicate that production polish may be less important than content relevance.

APPENDIXES

5.A: Democracy in Communication: Popular Video and Film in Latin America

by Karen Ranucci

Karen Ranucci served as project coordinator for *"Democracy in Communication: Popular Video and Film in Latin Amercia."* The following describes some of the films and videotapes that show this area of the world through the eyes of those who wish to make a statement largely from outside the traditional film and television establishment. Works included range from a five-minute tape depicting the life of an urban working woman to a feature showing Salvadoran guerrillas using videotape to communicate directly with peasants.

Latin America: Many North Americans read about it in the United States press, some write about it, while still others make films and tapes about it. For the most part information about our sister hemisphere comes from North American journalists and news agencies. Our perceptions of Latin America are shaped from our own cultural and political perspective. Rarely, unless one travels to Latin America, do we have an opportunity to know how Latin Americans express their own realities.

"Democracy In Communication: Popular Video and Film in Latin America" is a collection of works made by independent Latin media producers.

This project came about as a result of my international travels as a freelance journalist for North American television companies. In this work I observed how news and information are gathered and presented to the United States audience. Reporters often arrive in countries without speaking the language and with little or no background in the history of what they are sent to cover. The voice of the people is rarely heard. Usually only spokesmen who speak English appear on TV, as subtitles are difficult to use in news production.

After some time I tired of hearing our own voices. I spent a year traveling throughout Latin America, making contact with all of the independent producers I could find. The result is this collection of video tapes, shot and spoken by the peoples of Latin America.

KAREN RANUCCI
Project Coordinator

209

Many countries in Latin America have a long history of cinema production. However, the financial crisis which has wracked the continent has made it nearly impossible for independent film makers to work.

The video revolution, which began in the United States more than a decade ago, is now reaching Latin America. Many filmmakers have transferred their skills to video production and are creatively experimenting in a variety of community television projects.

Television is one of the strongest forces in forming the ideas and opinions of a population. Control of it means power. In most Latin American countries access to the airwaves for independents is nonexistent. To side-step the influence of broadcast TV, many communities produce their own television and show it in closed-circuit settings. "Popular" video is like an octopus reaching out in many different directions. In contexts where political repression has closed the channels of mass communication to democratic expression, alternative movements have arisen. In recent years, these isolated independent community groups have begun to share their experiences and create an alternative media network.

Throughout Latin America the use of portable home video cassette recorders is steadily increasing. Members of informal video clubs share pirated tapes and choose what they watch, rather than depending solely on broadcast TV. The cost of a privately owned VCR is prohibitive for most Latin Americans, but institutions such as churches, unions or collective groups can make the investment. See the chart which lists the use of TV sets and VCR's for each Latin country.

The democratic nature of video and other new communication technologies offers exciting possibilities for a direct link between the peoples of North and South America. The financial and political constraints, while limiting this potential, cannot stop the innovations made by determined communicators.

MEXICO

"Cross Section One Afternoon Of Mexican TV"
Video tape, 5 minutes, recorded off air, 1985

Mexico has both state and privately owned TV channels. The private channels are owned by the monopoly, Televisa. Televisa operates four channels, Channels 2, 4, 5, and 9. Fifty percent of its programming is imported, mostly from the United States. The majority of the program-

Country	Population (millions)	TV Set (millions)	People per TV Set	VCRs (thousands)	People per VCR (thousands)	% TV homes with VCR
Argentina	28.2	6.5	4.8	85	331.8	1.4
Bolivla	5.7	.36	15			
Brazil	120.5	19.3	6.2	550	219.1	3.3
Chile	11.3	2.0	5.65	65	173.8	2.5
Colombia	26.4	2.5	10.6	310	85.2	17.7
Ecuador	8.6	.4	21.5	40	215	10
Paraguay	3.1	.085	36.5			
Peru	17.1	.085	20	70	244.3	12.1
Uruguay	2.9	.36	8			
Venezuela	15.4	2.8	5.5	500	30.8	31.3
Regional Tot	239.2	35.155				

Source: Peter Ayreton, Tom Engelhardt and Vron Ware (eds.), World View 1985—An Economic and Political Yearbook. Pluto 1984; and World Programme prices and television statistics. TV World, June 1984.

ming on Channel 2 is produced in Mexico. This includes news, sports and old movies, as well as more than thirty soap operas and twenty variety shows per week. Channel 4 specializes in feature films and old United States TV serials such as Combat, The Addams Family, The Munsters and The Untouchables. Channel 5 concentrates on children's programming. One hundred percent of the cartoons are United States imports. Evening programming on Channel 5 consists of current American serials such as Magnum PI, Dallas, Falcon Crest, Dynasty, Quincy, Fame, Ripley's Believe It Or Not, and That's Incredible. Channel 9 is the educational and cultural channel. One of its most popular programs is Video Cosmos, which features American-made music videos.

As people have become accustomed to fast-paced American TV shows, they are often bored by the programming on the state television channels. It imports some of its programming, but runs many locally produced talk shows, science, historical, and educational programs.

Although this is a sample of Mexican TV, it is typical of that in many Latin American countries. That so much of Latin American TV is purchased from the United States has a definite social effect. United States ideals and values are exported and received by Latin Americans through their airwaves.

"Lugares Comunes" ("Common Places")
16 mm Film, 20 minutes, 1983

Mexico has had a rich history of independent film production. But the distribution of these works remains a serious problem. Only one theatre in Mexico, the Cinemateca, shows independent films on a regular basis, and only one distributor, Zafra, handles independent films. The high cost of film prints and lack of resources by groups who want to use these films have forced Zafra to distribute some of its films on video tape. Meanwhile, Mexico's two film schools continue to graduate hopeful producers into a world where there is no financing for independent film production

"Lugares Comunes" is an example of a student film written and directed by Lillian Liberman, a feminist filmmaker. It examines the reality of life for Mexican women by contrasting the stories of two women from different social classes. One of the women lives out her middle-class existence waiting for her husband to come home from work. The other is a young secretary from a working-class family who repeats the daily ritual of taking the bus to and from work. She comes to the realization that her only hope to escape this endless cycle is through marriage.

"Amas de Casa" ("Housewives")
Video tape, 5 minutes, 1984

"Amas de Casa" was produced by the Colectivo Cine Mujer, a group of women filmmakers who joined together in 1978 to produce programs which would be relevant to their interests as Mexican women. In this video tape project, women from one of the city's many cardboard villages act out a familiar situation. A neighbor, late with her rent payments, is being evicted. The entire neighborhood bands together to stand behind her in defiance of the court's eviction order. This tape has been used as an organizing tool by the housewives' union in this neighborhood. The process of making the tape helps the women gain the confidence they need to face this situation again, in real life.

Today the Colectivo Cine Mujer no longer exists. Rather, each of the women are working separately in some aspect of cinema production.

"Nuestro Tequio" ("Our Tequio")
Video tape, 10 minutes, 1984

The Zapotecas are indigenous people from the state of Oaxaca. Six years ago they formed their own video production group using Betamax ½" equipment. Four young Zapotecas work together to produce programs about their customs, institutions, and the political situation in their villages. They have already produced six programs and publish a newsletter about their work.

"Nuestro Tequio" is a video documentary about the ancient Indian custom of joining together one day per week to work on a community project. In Yalalag, the people have been working for three years to restore their city hall building. It is an important historical institution and is falling apart. No state funding was available to aid in the purchase of repair materials. The men of the community donated a portion of their income into a collective fund to buy what was needed.

This tape captures the day of the "Tequio," when hundreds of people from the region gathered together to put a new roof on the building. The tape was made so that other Indian groups could share the experience and to demonstrate how much can be accomplished by joint effort.

Says one of the tape's producers, "These Tequios are part of our history. When we see ourselves on TV doing this work, we feel proud. We also show the tape outside of our community to prove to others that we are not the dumb Indians some people believe us to be."

This production collective receives no outside financial support. They must work as migrant farm workers in the United States to earn the money they need to make their video tapes.

"El Triunfo" ("The Triumph")
Video tape, 15 minutes, 1985

"El Triunfo" is a tape made for environmentalists as a tool in their efforts to save the jungle from destruction. It examines the reasons for the destruction of the jungle and visits a jungle reserve which was created to protect endangered wildlife. It was produced by Video Servicios, a group of independent producers who have pooled their resources and talents to form a production company and community video center.

"Video Road"
Super 8 and Video tape, 10 minutes, 1985

In Mexico there is a small community of experimental video producers. Their work, for the most part, resembles early attempts by U.S. video artists. There are only one or two art galleries in Mexico where these works are shown. They are most often shown by the artists to friends at private screenings. This tape, "Video Road," by Sara Minter, is a documentation of a cross-country journey. It was filmed in super 8 and edited on video. Ms. Minter has her own video equipment and rents her services professionally, on a freelance basis. She is currently working on a series of projects about Mexican punk-rock street gangs.

BRAZIL

Of all Latin American countries the independent video community is the most developed in Brazil. ½" Portapaks were introduced there more than 15 years ago. Independents have been experimenting and building their skills since then.

Olhar Electronico is the most popular independent production group in Brazil. Their tapes consistently win first prize in national video festivals and are broadcast on television. Their productions included in this exhibition are:

"Marley Normal"
Video tape, 5 minutes, 1985

A day in the life of an urban working woman is condensed to five minutes in this experimental videotape.

"Varela In Serra Pelada"
Video tape, 3 minutes, 1984

Varela, a satirical caricature of a TV correspondent, directs this tongue-in-cheek report from Brazil's largest gold mining area. Thousands of prospectors work in mud-filled pits, hoping to find their fortunes.

"Varela In Xingu"
Video tape, 13 minutes, 1985

This satirical tape documents the inauguration of a new tribal chief on the Xingu Indian reserve in the Amazon jungle. The Indians give their impressions of White society, while Varela pokes fun at the orgy of network news crews who have arrived to cover this "media event."

"Sound On/Vision On"
Video tape, 19 minutes, 1985

This tape is a collage of Afro-Brazilian sounds and images. It contrasts these images with economic development projects that destroy Indian lands. It was produced by Enugbarjo, a community video collective that works closely with ethnic and women's groups. They shoot in VHS and have an archive of footage which relates to women and Blacks in Brazil.

"Beijo Ardente: Overdose"
Video tape, 40 minutes, 1984

In the city of Porto Alegre a group of artists unite to create a cultural center by reconverting an old gas plant. An independent video collective, Olho Magico, volunteered to make a video tape that would help gain community support for this project. They wrote a script that uses a vampire as a metaphor for the politicians and industrialists, who rather than using their resources to help the arts, suck the life blood from the artist community.

The tape won first prize at the Rio Video Festival and has been shown throughout the country. The artists won the right to create the cultural center, but the lack of financial resources has kept the project from being completed.

CHILE

In 1973 the democratically elected government of Salvador Allende Gossens was overthrown by General Augusto Pinochet in a military coup. Many years of brutal repression followed. Thousands were killed and "disappeared." All forms of mass communication were brought under military review.

The availability of video portapaks has spurred a movement where producers use video on a local level as a form of alternative communication. Through these tapes officially banned information reaches the populace.

"News Clips by Teleanalisis"
Video tape, 5 minutes, 1984

Teleanalisis is a group of freelance video producers who make alternative news reports. Their tapes have been outlawed by the censor and are forbidden to be shown in Chile.

This clip, "Jornada Por La Vida" ("Journey For Life"), documents the events which lead to a series of demonstrations by a broad coalition of organizations against the repression of the military government.

"Chile's Forbidden Dream"
Vido tape, 52 minutes, 1983

This program was co-produced by the British Broadcasting Company and ICTUS, Chile's oldest theater company. ICTUS adapts many of their theater pieces to video and has created an informal network to show these tapes in poor neighborhoods around the country.

"Chile's Forbidden Dream" provides an overview of Chile's recent history and places the work of ICTUS in this context.

"Hasta Vencer" ("Until Victory")
Video tape, 12 minutes, 1984

The housing crisis in Chile has become so great that thousands of once middle-class families have lost their homes. This documentary follows the birth and development of a squatter's village on the outskirts of Santiago, the capital city. It was made to show homeless people that it is possible to solve their housing problem by joining together with others in the same situation. It was produced by the ICTUS Theater group.

"Blanca Azucena" ("White Lily")
Video tape, 28 minutes, 1985

"Blanca Azucena" documents an experiment in popular education which took place in a mountain village in southern Chile. A group of peasants are trained to teach reading and crafts to other local residents. This program represents a form of "popular video" where the tape is integrated into a community setting. Through the process of making this tape, peasants are better able to critique their educational project. "It's like looking in the mirror and seeing yourself in a whole new way," said one of the program's participants. They hope their tape will provide a lesson and inspiration for other peasant villagers who may try a similar experiment.

PERU

In Peru, independent filmmakers have joined together to create a form of union. Grupo Chaski is a collective of more than 35 filmmakers which has produced a number of feature films and documentaries. Their objective is to create a national cinema which gives a voice to the marginal sectors of Peruvian society.

"Miss Universe in Peru"
16 mm Film, 28 minutes, 1984

This documentary was produced by Grupo Chaski. It is a behind-the-scenes look at the Miss Universe contest and examines its relationship to real life for women in Peru.

"Gregorio"
16 mm Film, 58 minutes, 1983

"Gregorio," also produced by Grupo Chaski, is a feature film that portrays the typical story of an Indian family who leaves the country-side to find work in the capital city.

The film shows the disintegration of the family by following Gregorio, the son of Indian migrants. After leaving his friends and family in the country, he is slowly transformed into one of the thousands of kids who roam the streets of Lima.

None of the young people in the film are professional actors. They are real street kids.

"Desparecidos" ("The Disappeared Ones")
Music Video, 7 minutes, 1985

Independent video production in Peru barely exists. The first National Video Festival, held in 1985, received a number of entries from producers who work for the state-controlled broadcast TV. One entry was "Desaparecidos," a music video of the song by Ruben Blades with images that document the situation in the state of Ayacucho. There, Indian peasants find themselves caught in a battle between the Peruvian Army and the Shining Path guerrillas. In recent years thousands of Indians have "disappeared" or been murdered, victims of the ever-expanding war.

PANAMA

"Algo de Ti" ("Something of You")
Music Video, 5 minutes, 1985

"Algo de Ti" won second prize in the 1984 Panamanian Music Video Festival. It represents the integration of music with social statement. The song is a love song, but the video speaks of the horrors of living under a military government.

A man is arrested with no explanation given. Surrealistic images imply that society lends its complicity through its silence about such acts of violence. It was made by Luis Franco and Sergio Cambefort of Boa Productions. They work in commercial video production and do their own work on the side.

BOLIVIA

In 1967, the Bolivian government began the country's first TV channel. Communication flourished in 1974 when a unique system of regional TV channels run by universities added eight more channels to the national broadcasting system. These university channels show many independent works.

The militant miner's union won the right to open a nationwide system of regional, popular radio stations. Since most rural areas in Bolivia do not have electricity, television is not valuable to communicate in the countryside. The union also created a super 8 film workshop through which miners, trained in film production, made films about their lives.

With the military coup in 1980, the university TV and miner's radio stations were closed.

"Lucho: Que Vive en el Pueblo"
("Lucho: You Live Within the People")
Video tape, 15 minutes, 1985

"Lucho" is a documentary about the life and death of Father Luis Espinal. He was one of the principal leaders in the movement for democratic communication and was murdered by the military government. This is a videotape tribute to the man and his work made by some of his film students.

In 1982 the country returned to democracy. The TV and radio

stations reopened. This tape was broadcast throughout the country and was seen by thousands.

Today, the makers of this tape, Alfredo Ovando and Liliana de la Quinta are the only people making independent video in Bolivia.

URUGUAY

"Señal de Ajuste" ("Signal to Adjust")
Video tape, 15 minutes, 1984

This tape is a comic adaptation of a short story in which a couple's relationship is ruined by their TV set. It begins with a monologue by the male protagonist. Accused of murdering his wife, he explains his story to the police interrogator. Acted-out flashback scenes show what happened.

To deal with the monotony of everyday married life, he bought a TV set for his wife. Little by little she became obsessed, integrating herself completely with the TV, until in the end she short-circuited.

The underlying theme of this story is the social alienation which is created by television.

"El Sol del Juez" ("The Judge's Coin")
Video tape, 10 minutes, 1983

"El Sol Del Juez" is an adaptation of an historical incident which led to the largest peasant uprising in Bolivia. One day a rich man unknowingly drops a coin on the street. The poor Indian villagers are dying to pick it up. Their history of oppression and fear cause them to longingly stare, but never touch the coin. The visual treatment of this simple story is like a video poem.

Both these tapes were made by Juan Jose Ravaioli of Estudio Imagen. In addition to making their own tapes, Estudio Imagen has a circulating video library of Latin films and tapes with more than 300 titles.

EL SALVADOR

The six-year-old civil war in El Salvador has been fought in the battlefield. But both the El Salvadoran military and the guerrilla forces understand the importance of the propaganda war.

"Atlacatl"
Video tape, 7 minutes, 1983

This is a publicity tape made by the El Salvadoran military about Atlacatl, the special forces brigade which received its training in the United States. The military's Press and Information Office makes propaganda tapes which are broadcast on the state-controlled television. These tapes serve in the military's campaign to gain popular support for their role in the ongoing conflict.

"Los Refugiados" ("The Refugees")
Video tape, 28 minutes, 1985

This program was not produced by Latin Americans. Its North American makers, Michael Ach and Mark Brady, worked closely with El Salvadoran refugees and religious support organizations. It was included in this exhibition because it gives illegal El Salvadoran refugees who live on Long Island, New York, an opportunity to explain why they left and to show the reality of their underground lives as refugees.

"Tiempo de Audacia" ("Time of Daring")
Video/Film, 22 minutes, 1983

The guerrillas feel they must communicate directly with the people of El Salvador to combat the government's control of the media and relate their side of the conflict to the populace. Their primary activity is in the form of daily radio broadcasts. In order to operate undetected by the military, their transponder is portable and constantly moving.

Their video work serves audiences both inside and outside the country. Their tapes are taken to mountain villages and shown to peasants as a way of keeping them in touch with the war that is being fought in other parts of the country. Their tapes are also shown in foreign countries as a means of gaining international support for their cause.

NICARAGUA

Under the regime of the dictator, Anastasio Somoza, no independent video was produced in Nicaragua. Since the revolution in 1979 many video experiments have been tried.

"La Virgen que Suda" ("The Sweating Virgin")
Video tape, 15 minutes, 1983

This program was also produced by SSTV. It is a made for TV drama that highlights the uneasy relationship between Nicaragua and the United States.

Uncle Sam is displeased with the cooperation he sees being formed between the Sandinista Government and the Catholic Church. He comes up with a plan to disrupt this alliance. He bribes a couple into going along with his scheme.

One day a statue of the Virgin Mary began to sweat. Its owners claimed that this "miracle" was a warning of God's displeasure with the Sandinista Government. It was later revealed that the owners soaked her in a bath and froze her overnight. This way she would appear to sweat when displayed with burning candles surrounding her.

This drama is based on a true incident.

"Testimonios" ("Testimonies")
Video tape, 15 minutes, 1982

The Timeteo Velazque workshop was organized by the country's largest trade union. It started as a super 8 class and quickly moved into video production. Their tapes are shown throughout the country and broadcast on national TV. This program, "Testimonios," describes the effects of raids by United States backed "Contras" on the lives of the Nicaraguan people.

"Las Mujeres" ("The Women")
Video tape, 8 minutes, 1985

This tape examines the disparity in pay between women and men on cooperative farms. It questions administration policy and shows a successful example of how that policy can be changed. The tape focuses on one farm where men joined together with their female co-workers to demand equal pay for all.

It was made by a women's video workshop and has been used as an organizing tool by women's groups in Nicaragua.

"Aquí en Esta Esquina" ("Here on This Corner")
Video tape, 5 minutes, 1985

"Here on This Corner" is a traveling game show produced by the Systema Sandinista de Television (SSTV), the official government television system. It broadcasts live from a different neighborhood each week. Local people compete in a variety of contests for money prizes, and others get the opportunity to display their talents. Some people have become famous as a result of appearing on this show. This clip features a children's salsa contest, a musical group called The Humble Ones, and an arm wrestling match between two women.

"Que Pasa con el Papel Higienico"
("What Happened to the Toilet Paper")
Video tape, 17 minutes, 1983

This tape was made by MIDINRA, the ministry charged with implementing agrarian reform measures ordered by the Sandinista Government. Their video unit has made a number of programs which examine the effects of the government's agricultural and economic policies. "What Happened to the Toilet Paper" is an investigative report which uncovers why there is a shortage of toilet paper in Nicaragua.

RESOURCES

Mexico

Distributor Independent Films and Videos:
Zafra
82 Leonardo DeVince
Mixcoac 03910
Mexico, DF, Mexico

Brazil

Guide to Independent Video Producers and Productions:
Do Guia Do Video, No Brasil
Olhar Electronico
Avenida Pedroso de Morais 1572
CEP 05420
São Paulo, SP Brasil

Chile

Guide to Chilean Independent Video Producers:
Video Independiente en Chile
Yessic Ulloa—CENECA

Santa Beatriz 106 Providencia
Santiago de Chile

Peru

Publisher:
Materials for Communication, a collection of articles and information about the state of alternative communication in Latin America.
New Communlcation Technologies, an international newsletter which serves as an organ for the infant network of alternative communication producers:
IPAL
Apartado 270031
Lima 27, Peru

Nicaragua

Distributor Videos from Nicaragua:
Exchange TV
Martha Wallner
P.O. Box 586
New York, N.Y. 10009
212–713–5544

Uruguay

Circulating Video Library
Estudio Imagen—Juan Jose Ravaioli
Av. Agraciada 1641, ESC 202
Montevideo, Uruguay

United States Distributors—Latin American Films/Tapes

Icarus
Jonathan Miller
200 Park Ave. South
New York, N.Y. 10003
212–674–3375

Cinema Guild
Gary Crowdus
1697 Broadway, Suite 802
New York, N.Y. 10019
212–246–5522

If you would like to show these tapes in your community, contact:
Karen Ranucci
656 Carroll Street
Brooklyn, New York 11215
718–499–9524

They can be packaged in any way that would suit your programming needs and are available in any format.

The rental fee charged for this exhibition is flexible. It is based on the amount of programming used and the ability of your organization to pay.

5.B: The Piracy of Cinematic Films on Videocassettes

by Henrique Gandelman
(translation by Elizabeth Senger)

Throughout this book, we have discussed the problem of video piracy and the impact it has had on the developing world. The following paper, "The Piracy of Cinematic Films on Videocassettes" by Henrique Gandelman, was translated by Elizabeth Senger and published by Union Brasileira de Video (Brazilian Video Union), Rio de Janeiro. (Copyright 1985 by Henrique Gandelman.) The essay discusses copyright problems faced by those attempting to show films in Brazil. However, Gandelman's discussion applies to virtually every country covered in this book.

> Authorship or piracy?
> —this is the question
> in the era of new technologies!
>
> —L. H. Myers

CHAPTER 1: WHAT IS PIRACY?

1.1 The Challenge of Technology

The violent and dramatic explosion of technology in this century is growing through the classic mediums of communication: radio, television, cassette tapes, videotapes, photocopying, satellite transmissions, telegram, and computer software. All this paraphernalia causes the media to compete with the actual messages, at times even supplanting them...(as observed by Marshall McLuhan!).

Each stage of technological development—paralleling the eventual progress of intellectual creativity that it promotes—also encourages its own types and characteristics of piracy. Even now the invention of machines that record and exhibit videocassettes—i.e., VCR, videocassette recorder, with the abilities it created—can turn any citizen into, involuntarily, the "author" of an illegal act: domestic piracy. And it is also an incentive not to go out to the movie theaters. This is the challenge of the videocassette!

The wish to communicate is, incontestably, at the beginning of each expression of creativity. In addition, one of the principal goals of technological progress that should always be repeated is the dissemination of culture, of entertainment, and of pleasure. The intellectual creations, independently of the communication mediums by which they are distributed—be it printed matter, audiovisual productions, or even a computer—should have their legal rights respected; therefore, returning in a dignified manner the respect for the talent and desire to create through which they were created. Without this respect, new works probably would not be created.

The following testimonies illustrate in a crystallized form what was previously observed:

What is happening now is the end of our synchronized society. Better said, the TV time slots are becoming more personalized. The television stations are programming to emit a uniform signal to all places, all the time. The videocassette lets you interrupt the synchronized massification of the television networks, freeing you from the concept of time (Alvin Toffler, interview in the magazine *Billboard*, New York, September 19, 1981, p. 52).

Now it is easier to watch all the films of the ancient cinema, or the current successes, in your own home. It's far simpler to record the programs of your preference, and operate your videocassette recorder.

In the recordings, when you want to eliminate a passage (commercials, for example) you utilize the key "PAUSE." The new videocassette X has an exclusive circuit that immediately goes into action, eliminating the "noise" present in the "revisions" of scenes, making possible a professional level of quality to this editing (taken from the bulletin "Nunca Foi Tao Facil," ("It Was Never So Easy"), published in the magazine *Veja*, on August 7, 1985, p. 152).

The expansion of audiences is one of the major contributions that technology brought to intellectual creation, which is already reaching a level of development unimaginable, as we said, a century ago, and promises in the near future to break all limits of numbers, of space, of time, until infinity (Georges Koumantos, "Challenges and Promises of the Mass Media for Copyright," in *PRS Yearbook*, London, 1981, p. 85).

Who never felt the gradual loss of a packed movie house over the

years either has a clouded memory or a heart of stone. (Carlos
Drummond de Andrade, in the article ["The Cinemas Are Closing"],
Jornal do Brasil, Caderno B, January 1, 1984.

The observations of Toffler (author of *Future Shock* and *The Third
Wave*), the public appeal by the makers of VCRs with reference to
technological capabilities to copy of watch feature films in the home,
the commentaries on the impact of the new technologies in mass
communication, as well as the exclamation of the poet are, without
doubt, a shout of warning for the future of the filmmaking industry.
This industry, besides threats due to social changes or an economic
recession, is now faced with, and directly threatened by, "electronic
piracy."

1.2 Piracy and Its Impact

Vulgarly it is called piracy, the act of copying or reproducing even for
purposes other than monetary gain, without the express authorization
of the copyright holders for books or other printed matter, video or
audio recordings, computer software, or any other physical product
that contains intellectual works legally protected.

It is not considered piracy, however, according to current legislation, to make a single copy in your own home for personal use. But if
this copy leaves the house to be reproduced, loaned, traded, shown
publicly, or in any other form utilized, without the express authorization of the copyright holders, this turns into piracy.

The start of piracy of cinematic films onto videocassettes was
caused by the following factors:

a. the technological versatility of the VCR, with which it
is possible to record or exhibit both image and sound,
plugged into a television set, even in your own home;
b. the easy money that such a "fantasy industry" can produce, since the cost to produce one illegal cassette is
cheaper than to produce the legal product, to say nothing
of the copyright fees and taxes saved! With only two VCR
machines it is possible to copy onto a blank videocassette
while one machine plays back a rented tape.

The obvious, yet false, conclusion is that in the era of the videocassette, any citizen is a film "producer" without having to pay anything
to legitimate producers, writers and playwrights, directors, authors,
musicians, cinematographers, technicians, etc. In truth, the struggle

against piracy, which is actually an international battle, is the defense of the right to authorship, the right of the community to create, and one of the fundamental human rights.

The parties touched by the illegal activities of piracy, and that directly suffer their impact are:

a. the country, the states, and the cities that allow this evasion of their respective taxes. With this, the entire community is also indirectly touched;

b. the businesses that produce feature films and that suffer great economic losses. In addition, when this interferes with the capital returned that can be applied to its legitimate productions, it becomes each time more difficult to promote new productions, or even to remunerate the creators (authors of the original texts, playwrights and adapters, composers and musicians, actors and interpreters in general, artistic directors of the films, etc.). With this, the legitimate distributors and exhibitors are also economically affected;

c. the public consumer, who is compelled to use copies of poor technical quality, with scenes cut and often dubious storylines—a veritable cultural mutilation.

d. the legitimate distributors of pre-taped videocassettes, which face entering a market already occupied, as well as the videoclubs and rental shops which use legitimate products, who all suffer the impact of a truly illegal competition.

Piracy today already costs the international cinematic industry per year around one billion dollars in lost copyright fees and governmental taxes. In Brazil there are no data about this yet. However, Embrafilme (Empresa Brasileira de Filmes S. A.) stated in 1980 there existed 2,826 movie theaters, while in 1984 only 1,549 cinemas were operating.... In 1980, 165 million people went to the movies, while already in 1984 only 90 million attended...(magazine *Isto E*, October 4, 1985, p. 51).

The commercial distribution of a feature film happens through a sequential series of channels, not in a monopolistic manner as with books and records, for example.

The three principle markets, in chronological order, from which the film industry obtains its financial returns are:

a. movie theaters, where the film is first launched, the first exhibition in 35 mm;

b. nontheater exhibitions—for example television by antenna, cable, or satellite, and in hospitals, hotels, ships, and airplanes, etc.—usually in 16 mm;

c. domestic video in the homes, through legally obtained videocassettes or videodiscs.

Each one of these markets can be a font or potential origin for the act of "electronic piracy" of videocassettes. This affects not only the market that at the moment is being legally patronized, but the next markets in the cycle are actually affected first, for already this form of piracy has destroyed the normal sequence of distribution markets: movie theater, nontheater exhibitions, and domestic video.

It is calculated that by the end of 1985, approximately 100 million VCRs will be in the hands of the international public. By this you can estimate the future of this market, which will be in the neighborhood of 500 to 800 million viewers.

1.3 The "Production" and Identification of Illegal Copies

The [Brazilian] domestic video market began at the end of the seventies, when tourists and visitors started bringing successful international films on videocassettes in their luggage to show on contraband equipment. Also, videocassettes could be bought on the black market in Manaus, as well as videocassettes already being produced within the country. Later, because of the novelty and increased demand, pirated copies of horrible quality appeared. International films, with an established audience in Rio, São Paulo, Brasilia, and other capitals, were copied as often as Brazilian films of *novelas* (serials) and musicals from television, which were more easily accepted by the less sophisticated consumers in rural areas. This caused an incredible proliferation of videoclubs and rental shops, creating a veritable boom in the industry, making it isolated from the general economic recession which had especially hit the film industry.

It is calculated that there are in circulation in the videoclubs and rental shops of the country between 600,000 and 800,000 taped videocassettes, the majority of an "irregular" origin. The "irregular" origin is probable since, according to the laws in force, it is not possible to import original videocassettes, but only the original filmed version whose copies must be produced in Brazil.

Even if the videocassettes were acquired outside the country by persons for their own personal use, here in the country they can not be copied, reproduced, loaned, traded in clubs, or shown publicly. This is to conform to the advertisements printed on the boxes of legitimate videocassettes acquired abroad. These inform the consumer of the protected copyright on the videocassette. This is a typical contract brought about by a simple purchase. which must be adhered to. In addition, copying is expressly restricted in the copyright laws, as we'll see later.

The videocassette reproductions not authorized are made through the following processes:

a.   transfer from film footage to videotape using special machines [film chains]; or taping an original film showing on television, creating an original film to produce whatever quantity desired of pirated videocassettes;

b.   reproduction of legal videocassettes left in the public domain through domestic equipment (VCR);

c.   direct recording by VCRs from television transmissions (normal, cable, or via satellite).

The pirated copies of videocassettes, in case of legal action, are identified by "**Intermedio de Laudo Pericial**," recognized by specialists. Therefore it is very easy for any legal consumer—videoclub associate, rental shop client, or purchaser of "pseudo-legitimate copies"—to soon ascertain if a videocassette is legitimate or pirated, by analyzing the following aspects:

External signs

a. absence of the seal provided by Embrafilme (Resolution of Concine #97/83);

b. generally the pirated copies have different **labels** than the originals and are contained in boxes from blank tapes of various brands acquired in the market;

c. the covers of the videocassette boxes are printed only with the name of the videoclub or rental shop, and have no information with regard to scenes from the film, title, names of the authors, the original producer, and the legal distributor.

Internal signs

a. the poor quality of the pirated copy, in both the color and clarity of the images as well as the sound;

b. the existence of subtitles, usually with grammatical errors, in works still not produced commercially within the country (because of this, they also do not bear the seal of Embrafilme);

c. the identification at the end of the videocassette of the name of the laboratory that transferred the film onto videotape, or of the television station that broadcast the film, since at times the logo of the original producer and other information of this type is cut from the beginning of the film.

We translated the following to illustrate this, a testimony realized by experts Drs. Newton Paulo Teixeira dos Santos and Luiz Rodrigues Romo, during a case presented in the 32nd Civil Court of the State of Rio de Janeiro, by various copyright holders against a videoclub:

> The experts could ascertain the following:
>
> *Gone with the Wind*. This film was entirely copied from a transmission by TV-Globo. The film itself is preceded by a short-length film titled *The History of "Gone with the Wind,"* with Metro (MGM) telling the first steps in the production of the movie, which was filmed in 1938–1939.
>
> It is inarguable that there was a "piratical" capture of the signals, and the defendant didn't even try to deny it. While the title of the program and the sponsors' advertising is shown, the voice of the announcer, from off camera, recites the commercials that supported the show. This was:
>
> Spring Festival.
> *Gone with the Wind . . .*
> Sponsored by Valisere—which unites hearts—Valisere—always a woman.
> Always have in the house drops of Anador, that reduce a fever quickly and safely.
> Onix Jeans, catch the style. See Onix.
> Milk of Aveia Davene—health for your skin for many years.
>
> Following this, the image of Cid Moreira appeared, famous announcer of TV-Globo, who announced: *"Gone with the Wind,* the film most famous and most watched in the history of the cinema. Box-office champion of all time. Earned the equivalent of $320,000,000." And recited a long passage that we won't repeat here, to introduce *The History of "Gone with the Wind."*
>
> After the Brazilian edition of this short-length film, the credits rolled, with all the characteristic signs of an interval on television. Next, the beginning of the main film was preceded by the image and words of actress Fernanda Montenegro presenting the film. It is inarguable that you are dealing with a pirated copy. The logo of TV-Globo and the signal that characterizes it appeared at every break. They were not at all preoccupied, as we already stated, with hiding the truth.

After this examination, the experts found out this film was dubbed by MGM-UA, the second author of this action, and given to TV-Globo Ltda. for the period August 14, 1983, to August 15, 1983, "exclusively for utilization on its television programs" (docs. Nos. 1, 2, and 3).

The Verdict. Comprising two cassettes, with an English soundtrack and Portuguese subtitles. Does not contain any advertising on the recording. The blank cassette (holding magnetic tape) is a standard type, with labels from VideoClube of Brazil printed on it. If it was an original cassette, it certainly would have the trademark of Twentieth-Century Fox, as follows:

20th
CENTURY
FOX
TWENTIETH
CENTURY–FOX
CORPORATION

As well as a credit roll like this: "A Zanuck/Brown Production of A Sidney Lumet Film, *The Verdict* stars Paul Newman, Charlotte Rampling, Jack Warden and James Mason. Produced by Richard D. Zanuck and David Brown, Sidney Lumet directs from a screenplay by David Memet based on the novel by Barry Reed. Burtt Harris is the executive producer."

In this case, the credits are indicated only on the film, at the end of which appears the following: "Copies made thanks to Internegativo of Lider Cine Laboratorio S. A." This is an actual reference to film copies.

Already through the absence of the seal of Embrafilme from the tapes distributed by the Club (doc. No. 4) and by the characteristics above described, you can agree the cassettes apprehended constitute "nonauthorized reproduction," which is the same as counterfeiting (art. No. 4, V, of Law 5988, on Dec. 14, 1973). And counterfeiting is the maximum violation of the Copyright Law:

"The maximum violation, ordinarily designated by the term counterfeit, also consists of a total nonauthorized reproduction of a work so protected" (Hermano Duval, [*Violation of Copyrights*], Rio de Janeiro, Borsoi 1968, p. 15).

And the specialist Duval added:

The lack of concern towards the offended author is secondary and could almost be considered a simple moral slight. Also not important is the external form the fraudulent reproduction takes, like a change of format or color, the reduction of the original to an abridged version, or its reproduction in a different medium (op. cit., p. 15).

Primum vivere, deinde philosophari (first live, then philosophize)—an expression popular in Latin that adapted well to our emerging market in domestic videos, until the present moment. It also reminds us of the good-humored advertising of a master author, done under the legal protection of intellectual values already consecrated in the civilized world.

When a furniture retailer buys a sofa from the manufacturer and pays the respective production cost, his only problem is to sell the sofa to a customer, preferably for a profit.

The administrator of a videoclub or rental shop or the distributor or seller of video cassettes quickly has the same problem, one of obtaining some profit. For him, however, exists another problem. He is dealing with a product that involves copyrights, which to him may be a complete mystery. In this case, it is definitely not a good solution to claim innocence of the Law.

1.4 Nonauthorized Uses

Nonauthorized uses of recorded videocassettes containing film works are usually made in the videoclubs, rental shops, or through public exhibitions:

a. the videoclubs—under the false concept that they don't make money, but "administered" with the final goal to produce profits for the owners—they accept from their members as many recorded videocassettes as blank cassettes, or even accept their monetary equivalent. This is besides paying a monthly membership fee, which supposedly is so the "administrators" can obtain videocassettes legally under the copyright laws, or "produce" their own copies;

b. the rental shops, that are in reality commercial firms, which rent in equal amounts videocassettes legally acquired, and videocassettes "produced" in the pirate laboratories;

c. the public exhibitions—in video salons, hotels, condominiums, etc.—to be considered legal, they must have previously obtained authorization from the Office of Censorship, Embrafilme, and the copyright holders for public exhibitions. Independent of any payment or ticket sales, the movie distribution companies are the only entities authorized for said activity.

The master Aurelio Buarque de Holanda teaches that the renter is one who in the lease, should be obligated to give up something, or offer a service (videoclubs?!!!), whether for a certain time or not. The regularity and eventual procurement of profit are essential factors that are always figured into a business transaction.

When the piratical videoclubs, rental shops, or public exhibitors utilize intellectual works legally protected, they're not acting because of affection, sentimentality, or the desire for cultural diffusion. They see their operation only as an instrument for making money, without any philanthropical considerations.

The illustrious judge Carvalho de Mendonca, in the Direct Commercial Treaty, (vol. 1, p. 472), classified commercial transactions as follows: "The purchase or trade of movable objects with the intention of: (a) reselling them, or (b) renting them; or still yet, the lease of movable objects previously bought or traded for this purpose." (Ask yourself: do the videoclubs promote the trade of things, with some goal of profit, directly or indirectly?)

Purchase again becomes an issue in the observations of Prof. Antonio Chaves published in the article ["Cassettes and Videocassettes"] in the newspaper *O Estado de São Paulo* (February 5, 1984, p. 47). In this article he traces with great propriety the legal problems the existence of videocassettes has caused:

> Connecting the theme of nonauthorized circulation of recorded material with a specific goal, we are forced to face as we produce examples of material produced without authorization, that there also exists "piracy" using legitimate material. This occurs when the material is used for a purpose not intended by the author when he gave his permission to use his work. This is, you can have "piracy" without new examples being created: a "circulation piracy" in which the material is legitimate, but it is not used for the purpose for which it was created.

The conclusion, therefore, is that, outside the boundaries of a particular residence, any use not previously authorized for either original videocassettes acquired abroad or those legitimately copied in Brazil, as well as those copies made for personal use, is a form of piracy.

Still on the subject of leasing or loaning of the physical products containing protected intellectual works, it is important to cite Judge Michael F. Flint, from *A User's Guide to Copyright* (Butterworths, London, 1979, p. 93), about the new English legislation:

The Public Lending Right Act of 1979 established a new right to be conferred upon the authors and recognized as "the right of public loaning or leasing"; which is not inherent, but related to the copyright law. The rights of authors are in actuality negative rights; that is, rights that impede others from exercising the exclusive transactions belonging to the owner of an intellectual work. The new right, therefore, is a positive right.

(Observation: see in 2.1.1, the decision offered by Dr. Pedro France, Judge of the 28th Civil Court in Rio de Janeiro).

CHAPTER 2: HOW CAN WE COMBAT PIRACY?

2.1 The Legal Protection

The Federal Constitution, in Article 153, Section 25, defines: "The authors of literary, artistic, and scientific works retain the exclusive right to use said works." Such protection, consecrated today in almost all the civilized countries, had its origin in the Universal Declaration of Human Rights, which stated in Article 27:

Every man has the right to the protection of his moral and material interests derived from any scientific, literary, or artistic production of which he is author (approved through Resolution of the Third Regular Session of the General Assembly of the United Nations, on December 10, 1948).

The most efficient way to combat piracy is through the recognition and acceptance of the entire community of the legal norms that determine the protection of intellectual works. To respect the law is, therefore, the starting point to any evaluation of the perspectives of a legitimate video market. The interest of everyone involved is fundamental, and shows the importance of cooperation between producers and copyright holders, judges, governmental authorities, retailers (videoclubs and rental shops), and consumers. From the pooling of forces, without doubt, will come the correct formula to guarantee the future of this activity.

2.1.1 The Connections with the Copyright Law

The scope of the copyright law has been behind the times, for the protection of the creations of the spirit—intellectual works—expressed in any form read, heard, or seen. This is a moral and legal guarantee of monetary retribution from society for the intellectual labor performed, in addition to organizing and distributing information, entertainment, and culture.

According to the philosopher Kant, this right would be one of exclusivity in which the author would have to authorize any public use of his intellectual work. Today, however, it is necessary to reevaluate this and adapt it to the brand-new frontiers created by the fantastic impact of the modern sophisticated technology of mass communication.

Historically, the limits not reached by the protection of the copyright law grew larger paralleling the technological development of the information media. After the revolution of Gutenberg, the book, the newspaper, and the magazine appeared, the culture imprinted on paper linear knowledge. Next, photography, radio, film, television, videorecorders, and satellites appeared, and the culture recorded audiovisually integrated knowledge.

We are living now with a technological explosion, in an era of collective work, of derived work, of piracy, of reproduction, and of utilization that is almost involuntary.

Only from the correct balance between the subject of this right (the copyright) and its object (the actual intellectual works), whose physical products in the stage of contemporary economic relations constitute important trading goods, can arise the actual legal protection for human creativity and for the existence of material possibilities for its continuance and permanence.

Brazil possesses today one of the most advanced laws in the world regarding copyrights, as the system is well structured and its amendment that refers to the legal problems caused by the modern communication media needed only small adjustments at the time we analyzed it.

We are referring to Law No. 5988, of December 14, 1973, which we give in the correct order as follows:

A. Article 1, Section 1—In this article it is stated that foreigners living abroad will enjoy the protection of international treaties and accords ratified by Brazil.

Bern Convention, revision of 1971 (Law No. 75.699 of May 6, 1975).

Universal Convention, revision of 1971 (Law No. 76.905 of December 24, 1975), concluding that, due to the force of the principle of reciprocation, the holders of copyright of both Brazilian and foreign films will have their works pro-

tected respectively in the territories where the same respects are paid, and whose countries adhere to the accords
referred to.

The Universal Accord in Geneva on copyrights, which the United
States and France signed, especially determines in Article No. 2 that
the published works in the United States enjoy total protection in
France, the same protection French works obtain under French law.
Therefore, it is the French law—Law May 11, 1975—that should be
applied to this: foreign authors or their agents benefit in France from
the French law, even if the use of their work is free in the original
country where the edition was produced and fabricated. And the
cited law, in particular Article 31, line 3, accepts the rule of territoriality of copyrights, and more expressly the rule of nonlimitation of
the domain of the exploration of the given rights, as well as for its
extension, destination, place and duration. (Decision of the 17th
House of the Grand Tribunal Instancia of Paris on April 24, 1980,
published in *RIDA* [*Revue International du Droit d'Auteur*], No. 106,
October 1980, p. 136, author Re Beatrice G.)

In this respect, the principle of territoriality, usually part of copyright
laws, plays an important role. Everyone knows that the copyright
regulation of a country only serves as a base and a structure by which
the actual legal situation in the country can be verified. The holder of
the right of distribution in Swiss territory, for example, will have his
rights determined by Swiss law. The holder therefore enjoys all the
specified advantages of this legislation as the distributor of a compact
disc, record, work of art, or book. The same person can be the holder
of distribution rights in another country—for example, the United
States. These rights, as we already explained, will be completely
different in content from the rights established in Switzerland, as
now the law of the United States will be applicable. The extinction of
the right of distribution applicable to the American example, after the
first sale, does not affect the rights of distribution in Switzerland,
one of the times the Swiss law does not contain any reference
to this. Next, the American sample, that eventually can circulate
freely in the States, will be submitted in Swiss territory to Swiss
legislation, in which it is explicit that a work cannot be "put out for
sale, lease, loan, or any public mode of distribution"—it cannot be
shown publicly, that is, for publicity goals, without the express
authorization of the holder of the distribution rights (public exhibition) in Switzerland. It can then be charged for this rule, since the
Swiss legislation does not acknowledge the fact that the sample was
legally imported. A simple public offering, made in Switzerland, to
buy, rent, etc., a sample of a foreign compact disc, would be contrary
to Swiss legislation, unless the holder of the distribution rights in
Switzerland had conceded the authorization for such a use.
The Switzerland/United States situation is here exemplified in such

a way as to demonstrate the character of the so-called first sale doctrine. In addition, it shows the character of the rules governing the extinction of the rights of distribution in a market, which can be "neutralized" when the samples that fall under the aforementioned doctrine penetrate another country, where a different philosophy concerning the rights of distribution of physical products containing intellectual works is practiced (Prof. Dr. Gunnar Karnell, "Rental and Related Market Phenomena Concerning Videograms and the Right of Distribution under Copyright Law," published in *RIDA* [*Revue Internationale du Droit d'Auteur*], No. 115, Paris, January 1983, p. 103).

In Brazil, however, which National Law No. 5.998/73 refers to, the Tribunals of State Justice are responsible for judging whether the respective civil actions already existing, to this respect, are sufficient and of definitive jurisprudence. "It is left to the State Justices to determine action in which copyright law is involved, consistent with internal legislation, even though it may conflict with international treaties" (5th Civil House of TJRJ, Agravo of Instrument No. 5720, unanimous decision, D.O. of September 20, 1982, fls. 106/110).

Because of the attention the above decision warranted, we also noted the following passage: "The hypothesis we used is identical to those resolved by the E. Supreme Federal Court: the legal precedent of action is not prescribed by the International Treaty or Contract; instead, by the Law No. 5.988 of December 14, 1973, which regulates the copyrights of this country, internal legislation."

We also wish to add, likewise, that the decisions of the E. Supreme Federal Court mentioned and reaffirmed by the E. 5th Civil House, are a confirmation of what is stated in Article 10 of Law No. 5010 of May 30, 1966, that established the jurisdiction of the Federal Justice.

The principle of territoriality is also confirmed in Law No. 5988/73 in Article 53, Section 2, in addition to what was already embued in the body of Article 1, section 1, where it reads: 'The following will make up the instrument of judicial powers, specifically the cessation of objective rights, the conditions for its exercise referring to time and place, and if it was a grave offense, refer to the price and retribution."

This signifies that the contracts dealing with the end of copyrights, should contain expressly the territories in which an annulment will be negotiated, since it is the law of these territories that will be competent to protect the intellectual works that were the object of the accord.

B. Article 4, VIII—This article defines, and meanwhile, protects the videocassette recorder (VCR).

C. Article 4, X, a and b—Characterization of the producer of videocassette recorders and films.

D. Article 6, VI—The film works and analogs are intellectual works expressly protected.

E. Article 15—The recognition of the legal person or company (*empresa*) who is holder of the copyright.

It will be retained for the author, in the condition of the realized production of a collective work, the right to utilize and benefit from the work of art he produced, in keeping with a vigilant legislation (7th Civil House TJRJ, Amendment No. 23.797, unanimous decision, October 19, 1982).

F. Article 29—The exclusive right of the copyright holder to utilize, benefit from, and allocate his work, as well as authorizing a third party to make it.

G. Article 30, IV, d—The express authorization of the copyright holder is always necessary prior to any form of utilization.

H. Article 35—The diverse forms of utilization are independent among themselves, therefore they should be separately authorized (for example, a writer's authorization for his novel to be adapted to a film does not mean that he has authorized the use of his work in the form of a *telenovela* (serial) on television; in the same vein, to use a film work in video it is necessary to obtain the express authorization of the copyright holder for both the transfer of the physical product and the public exhibition).

I. Article 37—The producer is the administrator of the patrimonial rights of a film.

J. Article 38—The purchase of a legitimate sample of a work—that is, the physical product—does not signify that, without express authorization, the purchaser can use it for whatever he wants, with or without the intent of making a profit.

Law No. 5988/73 determines some limitations—Article 49—and these limitations are also internationally recognized and called fair use, or justified use. Such restrictions, however, should be understood only as the intention of the legislators to not permit the interruption of the flow of ideas and information so necessary in the contemporary world, and not as a way to compromise the exclusive rights of the copyright holder to authorize the use of his works.

The copyright law of West Germany includes in one section around eighteen articles referring to fair use.

They include exceptions to the exclusive rights of the holders, such as the use in judicial proceedings, religious and education topics, reports on current themes, private use without the goal of profits, demonstrations to clients with a goal of demonstration of products, citation in a larger work, newspaper articles, etc. The same can be said of the Japanese legislation (1970 law), for example (Edward W. Ploman and L. Clark Hamilton, *Copyright—Intellectual Property in the Information Age*, Routledge and Kegan Paul, London, 1980, p. 86).

It is important to reassert that the monopoly of the exterior commerce of the Soviet State only permits the importation and sales in the USSR of any copy of works manufactured in the exterior, if such copies were acquired by organizations that operate legally in the cultural areas. The existing legislation in the Soviet Union also guarantees the authenticity of all the copies of the works that enter the domestic market, independently of technical mediums and physical products that these works present. This is reason for which, when they acquire cassette tapes, audiovisual tapes, or printed editions from the exterior, the Soviet organizations require, by contract, that the retailers guarantee the legality of the products that are being negotiated. This signifies that pirated copies do not penetrate the markets in the USSR. On the other side, and agreeing with Soviet legislation, to copy, duplicate, or reproduce works for commercial ends is not permitted to any citizen, individually. (From the announcement of G. L. Kolokolov, Director of the Agency of Copyrights of the USSR, presented in a forum about piracy, held in Geneva on March 16–18, 1983, under the patronage of WIPO [World Intellectual Protection Organization].

K. Articles 87 and 91—in these articles the obligation of the producers of film and its analogs are established, as well as those of the participants of their creations (and from these you can also calculate the damages caused by piracy.

In addition to these articles, we noted the following articles from Law No. 5.988/73:

Article 122—Whosoever prints a scientific, artistic, or literary work without authorization of the author, will lose any copies that are apprehended in the author's custody, and pay the author for the rest of the editions at the price at which they were sold, or were available for.

If the number of copies that make up a fraudulent edition is unknown, the transgressor will pay the equivalent value of 2,000 copies, in addition to any remaining copies.

Liquidation by articles. When dealing with indemnification for copyright fraud, based on Article 122, single section, of Law No. 5.988 of 1973, the 2,000 copies should have their value calculated by the price of the last sale, to remain in accord with the law which provided recourse (unanimous decision of the 5th House of TJRJ, Civil Appeal No. 30.522, D.O. August 1, 1984).

> Article 123—The author, whose work was fraudulently repro-
> duced, divulged, or in any form utilized, can, and as much as he
> wants, require the apprehension of the reproduced copies or the
> suspension of the divulgence or utilization of the work, without
> penalty to the right to indemnification of all losses.

In this case, the civil action would start with a Cautionary Measure
to Search and Apprehend (with a preliminary request)—Article 842,
Section 3, of the Civil Process Code. The finality of the measure is the
attainment of material proof of the illicit act, and the surprise of the
action is to get rid of the threat in the destruction of this proof, since
videocassettes can easily be erased or removed to another location not
within the police district. And thirty days after the activation of the
arrest, it goes back to the beginning, depending on the ordinary action,
which cause the verification of indemnification of losses—Articles 122
and 123 mentioned above.

In reference to the registration of copyrights, it must be observed
that in our legislation such measures are only declared, and not attribu-
tive. Juris Tantum constitutes a supposition since until it is proved
otherwise, the copyright holder of an intellectual work is a physical or
legal person who identifies this is his condition, in his own use of the
work—Article 12/16 of Law No. 5.988/73.

From what is seen, it can be affirmed that the best security for
unedited intellectual works (those that have not yet been published) is
the registry which only proves its anterior position, date of creation, or
the end of patrimonial rights—Articles 17/20 and 53, Section 1 of Law
No. 5.988/73.

This is the notion of an international register collected by the
Universal Convention already mentioned—seen in Paris (July 24,
1971), and which Brazil acquired (Law No. 76.905 of December 24,
1975). In Article III it defines the nonessentiality of the formalities to
protect the copyrights, in publishing the first edition of an intellectual
work, a notice that contains the name of the copyright holder and the
year of publication of the work, preceded by the copyright symbol is
sufficient to obtain the copyrights.

The videocassettes acquired abroad as well as those legitimately
produced in Brazil contain imprinted on them the copyright symbol,
this being sufficient and indisputable through the interpretation of
the law dealing with the manner of registering, conforming to the
previously mentioned method:

> The registration of a work is a declaration that constitutes the
> copyright (4th Civil Court TJRJ, Appeal No. 32.189, unanimous deci-
> sion, February 1, 1985).

Work of art. Facilitative registration not obligatory. Legal Protection. Proved the author of creation, then falsification or use for lucrative goals, by documents, testimonies and expertise, comes under the legal indemnification, inclusive of moral outrage (2nd Civil Court of TJRJ, Appeal No. 15.823, unanimous decision, April, 2, 1981).

The company Cic Video Ltda., for example, states in their videocassettes besides the copyright mark the following:

> Warning: The copyright holder of this film work—including its soundtrack—contained on this videocassette, is authorized only for your private and domestic use. Without permission and express authorization of the copyright holder, any other form of use is prohibited, such as copying, editing, adding, reducing, exhibiting or diffusing publicly, emitting or transmitting by radiodiffusion cable, or any other communication medium already existing or that will be created; as well as trading, loaning, renting, or practicing any other act of commercialization for direct or indirect profit. The violation of any of these exclusive rights of the copyright holder, will face the sanction laid out in Law No. 5988 of December 14, 1973, and Articles 184 and 186 of the Penal Code (Law No. 6895 of December 17, 1980).

The number of VCRs in public hands increases every day, as was already observed. The comfort of private viewing, the domestic pleasure among friends, the ease of use—everything therefore will contribute to the rapid expansion of this new medium of communication, which, as it seems now, really has a brilliant future (despite present difficulties).

Today in Brazil there exist about 700 videoclubs and rental shops in full operation. But in their showcases, the majority of the copies offered to the public still consist of illegal copies, not authorized by the copyright holders, and therefore, pirated works.

The transition of this new market to normalization has provoked various demands on the existence of the copyright law. There already have been positive decisions of the part of the judicial power—decisions like the pioneering sentence proffered (on August 8, 1985) by Dr. Pedro Arruda Pinto of France, judge of the 28th Civil Court of Rio de Janeiro, which is transcribed below:

> *Decided*: Following the declared action arranged in Article 4, I, and single paragraph of the Civil Process Code, belonging to the authors, film producers established and seated in the United States of North America, that as sentence it has been declared that "the lease on the

part of the leaser of videocassettes, reproducing of work of cinematography that belongs to them, without their express authorization, constitutes a violation of the copyright laws, conforming to Articles 29, 38, and 123 of Law No. 5988/73.

In the trial, the verification of the facts narrated in the beginning we will consider the negation of the violation of the copyright law of the first degree, seen by the Brazilian Association of Videocassette Distributors (ABDVC), made in contention and profiled by the latter to the specific legislation, indispensable is the force of the judicial branch.

Register that the legislator of 1973 did not introduce into the law the expression videocassettes but that, conforming to the parts the litigants situated themselves in, corresponds and is referred to inside the part that mentions videodisc/tape, as if it was an extension from the judicial concept.

Following the constitutional text, inserted in Articles 152, Section 25 of C.F., as well as Article 1, Section 1, of the law of civil proceedings these protect the copyrights, and those that are connected through accords, conventions, and treaties ratified by Brazil. In theory, the right invoked by the authors warrants the challenge of national law, by reason of reciprocity. This aspect is passive in the proceedings. Instituted at the Universal Convention signed in Geneva on September 6, 1952, and revised in Paris on July 24, 1971, and consecrated into law by Decree No. 76.905/75.

The videocassette is a tape with a recording of a film. This has its own author who warrants legal protection. To record on a tape called cassette is to reproduce, without sophistry.

The recording without authorization of the author of the work is counterfeiting according to the law. The producer of the film, a cinematic activity, is obviously the author of a work, on which he has worked proficiently, ardently, and which can not be undervalued by reproductions in copies not authorized by the author, a legal or physical person. The counterfeiting is described in Inc. V. of Article 4 of Law No. 5988/73 and in Inc. X, a and b; a record or videorecord producer that produces a record or videotape for the first time; and the cinematographer, with the responsibility of making the work a screen projection, establishes authorship. A cinematic work is of the order intellectual (Article 6, VII) and is holder of the patrimonial rights under it (Article 21) and thus holds the right to use, benefit, and distribution of the work as well as the right to authorize its use or benefit to third parties, for communication to the public, by any form or process (Articles 29 and 30, IV, d) of videorecording.

And if the first defendant acquired, by way of importation, the original videorecordings, conforming allegedly in response, with legalization of the monetary tributes, such a circumstance would not remove the illegal act of counterfeiting, an offense against the rights of the authors, producers of the films, because their authorization cannot be implicit for reproduction of leasing of the tapes.

It is true that the complaint of the authors against the first defendant, is that of damages experienced by the leasing of the videocassettes, and that this violates the express right of the law referred to above.

There is no doubt that the lease is a form of commercialization, following the first defendant *animus lucrandi*. The lease implies the end of use (the *jus fruendi*) of the owner of the work (or better said, the copyright holder), and that by not having their previous authorization, you enter into violation of Article 29 of the law under discussion. The leaser is no longer directly using the work of the author, without his consent, in a firm contract with which no one could disperse that use without that authorization.

For these reasons, the authors are authorized to exercise the right of constriction on the first defendant or infractor of the law that used artificial means in the practice of an illicit act, as previously stated in Article 123 of Law No. 5.988/73, as typically manifested.

In light of this explanation, proceeding to judge the act and declare that the lease on part of the VHS Video Cassete Rent Clube Ltda., an activity discovered by the Associacão Brasileira de Distribuidores de Video Cassetes (ABDVC), monitor of the club and that outlined the reasons for this suit, of videocassettes containing cinematic works owned by the authors Walt Disney Productions and Universal City Studios Inc., without the express authorization of the same, constitutes violation of their copyrights, conforming to the orders of Articles 29, 38, and 123 of Law No. 5.988/73, and Article 4, I, and the single section of the Civil Process Code.

Losses and their good word dubiously corrected in 20 percent above the value attributed to the cause, in proportion by the defendants, ex-vi of Article 20, Section 3, combining with Articles 23 and 62 of the Civil Process Code.

2.1.2 The Provisions of the Penal Code

Piracy is a specific crime in Articles 184 and 86:

Article 184. To violate copyrights:
Penalty—Three months to one year imprisonment or fine.
Paragraph 1—If the violation consisted of reproduction, by any means, of an intellectual work, without express authorization of the author or of whoever represents him, or consisted of the reproduction of recordings and vidoerecordings, without authorization of the producer or whoever represents him:
Penalty—One to four year imprisonment or fine.
Paragraph 2—The same penalty as above will be incurred by those

who sell, put out to sell, introduce to the country, acquire, hide, or have on deposit, to sell, the original or copy of an intellectual work, record, or videorecord produced with violation of the copyrights.

Article 186. On the crimes covered in this chapter only one proceeds by means of complaints, except when practiced in damages to the entity of public, governmental company, public company, society of mixed economy, or foundation instituted by the public power, and in the previous cases in paragraphs 1 and 2 of Article 184 of this law.

Analyzing the articles cited above, you can conclude

a. Paragraph 1 of Article 184, "to affirm—or consist...," is referring, evidently, to whatever means of utilization outside the home of videocassettes "reproduced" without authorization of their producer, or whoever represents them;

b. Paragraph 2 of Article 184 stated that any "administrator" of a videoclub or rental shop—this is all those who acquire, hide, or have on deposit physical products containing works "produced" in violation of the copyrights is subject to the previous penalty, already dedicated to the practice of commercial acts in the definition of Carvalho Mendonca referred to in 1.4.

c. The crime of violation of the copyrights is of public action.

It also should be mentioned in the legal struggle against piracy, one may apply Articles 171 (swindles), 180 (reception) and 196 (illegal competition).

With relation to the problems of penal suppression against the violation of the copyrights, Antonio Chaves made the following comment in a published article in *O Estado de São Paulo*, June 7, 1981, p. 58:

But, lessening the infraction, we would not have more to guard as before, the initiative of the interested and his option between the civil road and the criminal road: diligence, in the majority of cases, entails police authority, with the already analyzed amplitude of attributions, another prerogative exclusive of the Judicial Power.

And later concluded, referring to the withdrawal of the complaint:

The fundamental alteration will consist in that the first provisions will have its start with the opening of inquiry, ending with the police authority, the naming of experts and determination of warrants and arrests.

2.1.3 Industrial Property: Copyright
Trademarks and Logotypes

The judicial protection by means of legislation of industrial property, presupposes that the videocassette illegally reproduced contains on its body a trademark or logotype of the cinematic producer—that the beginning or end of the tape, or that the box in which it is packaged is printed with such symbols of the producer and/or the authorized distributor. Without such conditions, this protection is futile.

The law that regulates within the country this matter—No. 5772 on December 21, 1971, called the Industrial Property Code—determines that Article 175 of Decree-Law No. 7.903 on August 27, 1945, continues strongly and, meanwhile, makes the registered trademark or logotype necessary for any revindication of its ownership.

Article 175. To violate the industrial or commercial trademark:

I. Reproducing, improperly, all or in part, another's registered trade mark, or imitating it in a manner which could cause error or confusion.
II. Using the trademark reproduced or imitated in the terms of number I.
III. Using the legitimate trademark of another on a product or article not of your production.
IV. Selling, putting up for sale, or having on hand:
    a. an article or product with a trademark abusively imitated or reproduced in total or in part;
    b. an article or product that bears another's trademark and is not of your manufacture.

Penalty—Three months to one year imprisonment, and a fine of 1,000 to 15,000 cruzeiros.

Article 178 of Decree Law above cited the treatment of crimes of illegal doings, but the problems persist, as in the case when the pirated videocassette or its box lacks the trademark or logotype of its manufacturer.

It is very clear, in the conditions described, the protection of Law No. 5988/73 (Copyright Law) is more inclusive and adequate under the technical prism, to combat this form of piracy.

As observed by Walter Moraes:

Resembling photographical production as an operational and industrial process, and in the sense of producing a work of the spirit, is the production cinematic.

From the point of view of intellectual rights it is, after all, a business to generate products, as it has as its objective to market a work multiplied by copies: the producer acquires the copyrights to reproduce and sell a work, and explore such productions for sale and lease of copies ("Artistas Interpretes e Executantes," *Edit. Revista dos Tribunais*, SP, 1976, p. 263).

2.2 Other Forms of Protection

The film industry can not pass by this important segment of distribution of its products—called home video—inclusive to guarantee its very existence. A legitimate market can bring, consequently, an enormous impulse to the Brazilian economy: an increase to the collection of taxes, new jobs with the creation of laboratories and the necessity of specialists—including also the development of the correct techniques of marketing—the major interest of the factories to better the quality of copies offered, etc.

To talk of the subject of public interest—besides its cultural connotations—other forms of combating piracy are becoming permanent. Research, principally in that which refers to the physical aspects of the videocassettes, administrative resolutions of governmental bodies, and owners' associations realizing more clearly the state of public opinion has confirmed what we see the future as bringing.

2.2.1 The Mechanisms of Antipiracy Security

The efforts in the creation of strategies and technological mechanisms against piracy developed by the film industry, soon should include the moral integrity of their employees, managers, and collaborators. A film is placed at public disposal through various types of showings, and there exist many areas of vulnerability during the entire sequence distribution. One dishonest employee in the production studios, in a laboratory for developing or copying, in the film-transportation companies, or even among the theater employees is often the first responsible for acts of piracy.

In regards to the mechanisms of security, of great interest is the observation of Beatrice Tarouca:

Besides combating the pirates through judicial proceedings, the industry is increasingly turning to sophisticated means of prevention against the illegal appropriations of its products. The perfection already attained by the pirates in the areas of reproduction of the artistic elements of the box and seals of the videocassettes makes it difficult, even for specialists, to distinguish a pirated copy from a legitimate copy. Therefore, it is essential to prevent a pirate from counterfeiting, to make it easy for Justice Officials and Experts to recognize a suspect product in the stands of the videoclubs or renters, on the occasion of a warrant and arrest. There have been various cases where it was necessary to return the confiscated tape because of the difficulty in proving its illegitimate origin. Thus the reason to develop security measures is to make tapes easily identifiable. This measure consists of various systems of marking the legitimate product to permit its identification. These tapes are already in heavy use by industrial companies of records and videotapes/discs. The attempts to reproduce the security seals, it must be said, have not met with success.

There are a great number of security systems on the market. The one most utilized by the industrials is the security seal affixed and definitively incorporated into the box that contains the videocassette. Other security measures utilize papers specially prepared for the printing of the outer boxes, or stamped with holograph seals that can be numerically coded, and that also allow them to easily control production and the quantities produced.

The British Videotape Association is presently examining an electronic system—watermark—by which a code is included in the tape through a decoder: if the tape doesn't have the code, the decoder gives a signal that the tape is illegitimate ("Video Piracy: Fighting Back," in *International Media Law*, ed. Oyez Longmann, London, October 1983, p. 125).

It is important to realize that one of the effects of the physical marking of the legitimate videocassettes with security signals is to no longer permit pirates to allege good faith—i.e., no knowledge that they were in possession of a nonauthorized product.

2.2.2 CONCINE and the Governmental Agencies

The governmental support of the cultural industry is of extreme importance for its development, a factor that is already incorporated today into the executive politics of various countries.

The National Council of Cinema (CONCINE) even now, seeing the importance of the new market, has just finished creating two chairpersons for video in its plans (Portaria of the Minister of Culture, published in the *Diario Official* of August 21, 1984). One of these was destined for the owners of the movie rights and the distribution rights for video, and the other, to the factories of the commercialization of videocassettes.

Existing strongly at the moment are four resolutions of CONCINE involving aspects of the video market, Articles 97, 98, 99, and 106.

Probably, with the cooperation of the two new chairpersons in the plenario, these will be consolidated in the near future, with visions of their perfection and practical adaptation for their goals.

The competence of CONCINE to create resolutions is fundamentally in the insertions II, III, and V of Article 2, and in the form of Article 8 of Decree No. 77.299 of March 16, 1976, that regulated Law No. 6.281 of December 9, 1975. Others considering the resolutions of CONCINE affirmed:

> Considering that there is no legal distinction between cinematic works and the techniques of reproduction, the legislation, copies produced or copied by conventional processes compared to those that were electronically made, as it stated in Insertion IV of Article 6 of Law No. 5.988 of December 14, 1973, and the only paragraph of Article 1 of Decree-Law No. 1.900 of December 21, 1981.
>
> Considering that the videocassette is a magnetic tape placed in a plastic box of conventional size, capable of recording and reproducing cinematic work.
>
> Considering the directors of common action, relative to the subject, fixed by representatives of the Director of External Commerce of Banco do Brasil S.A., of the Secretary of the Federal Reserve, of the Division of Censure and Public Diversions, and the National Council of Copyrights, and of the Brazilian Film Industry (Embrafilme).
>
> Considering that the vigorous judicial order, at the disposal about the politics of cinematography, has as its objective the promotion of measures that are necessary to protect the development of the national industry.
>
> Considering that the increased use of cinematic works electronically recorded on videocassettes makes urgent the adoption of mechanisms that assure the participation of the Brazilian producer in this mode of information diffusion.
>
> Considering that cinematic work is the prior register of images in movement, independently of the technology used and the genre expressing it, presented in any pattern and in any system, recorded or reproduced on film, tape, videocassette, or still yet video disc, "video tape" or any other method of recording and reproducing sound and image, to exhibit in the theater, television, or any other vehicle under the terms of the only paragraph of Article 1 of Decree-Law No. 1.900/81.

The Resolution 97/83 of CONCINE created a security seal that would constitute a real guarantee for the easy identification of the legal product.

The principal topics of this resolution are:

a. Only to be commercialized in the country, for public or private exhibition, will be portable videocassettes with specially numbered labels, acquired from Embrafilme.
b. Only videocassettes in Brazilian territory copied from masters definitely registered can receive the control label.
c. Only the masters of Brazilian films or foreign films legally imported can be registered, when are proven, respectively, the copyrights, or the rights of distribution and commercialization for video.
d. Subject to warrant and arrest all the videocassettes in any place found that do not carry the control seals, or whose number does not correlate with the one on register with Embrafilme with the masters from which it originated.

About the external characteristics of legitimate videocassettes, Law 6,800 of June 25, 1980, still must be mentioned, which will determine that videodiscs/tapes that can not be sold, or maintained on deposit for commercial activities, unless they bear prominently, integrated into an indissociable form, the number of inscription of the General Register of Contributors (CGC of the Minister of Agriculture) of the industry responsible for the industrial process of reproducing a recording.

2.2.3 The UBV and the Association of the Owners

The UBV (Brazilian Video Union) is a nonprofit civil organization whose objective is to coordinate the defense of the legitimate interests of its associates, acting in the following areas:

a. Support and assistance for judicial actions proposed individually by its associates, in the sense of confirming the Brazilian legislation and international conventions about the violation of copyrights;
b. Assessor of the creation of antipiracy campaigns, informing the videoclubs, rental shops, and the public in general about the principles that rule the legitimacy of the market;
c. Providing representation to be present in the public authorities at all levels, including all the official regulated agencies of materials, in the sense to safeguard the rights of its associates, referring to both already existent legislation and elaboration of new laws, norms, and resolutions, relative to the distribution to the public and the commercialization of videocassettes.

Any legal or physical person can be an associate of UBV, as long as they conform to the following statutes:

a. You must be the holder of a copyright of cinematic works, or those produced by any analog process of that of cinematography, that has already been shown for commercial purposes.

b. You must be the holder of the rights of commercialization of videodiscs/tapes containing intellectual works protected by Law No. 5.988 of December 14, 1973.

It must be understood that the UBV is not an association for the protection of copyrights; therefore, it is not subject to the rules of Articles 103, 104, and 105 of the law referred to above.

The associations for copyright holders are important and decisive tools and strategies to combat piracy; and the international experience has proved that, with a well-coordinated activity, it is possible to achieve the existence of a legitimate video market.

The great technological revolutions through which humanity has passed always permitted the appearance of its creative geniuses. Every epoch is characterized by the culture it produces, by the creativity of its thinkers, artists, scientists, and legal thinkers. It is necessary, therefore, to ensure the intellectual right, the only ethical and legitimate form that legally is defined as propriety, in the epistemological sense of belonging to the good, uniquely, to those who created it. Without this we will no longer have more new creations, nor geniuses, in the society of tomorrow.

CHAPTER 6

Videocassette Recorders in the Caribbean

The Caribbean represents in a microcosm many of the characteristics, trends, and problems of the Third World. The region has a diversity of languages, cultures, geographical configurations, and political and economic systems.

Very few regions of the world have been more victimized by European imperialist policy than the Caribbean. For years, some islands seemed to be pawns in the political games of Europe; in the process, St. Lucia flew under thirteen different flags, St. Croix, seven. Such superimpositions of European culture left the people confused as to their roots and their sense of West Indianness. As Eric Williams, the late premier and scholar of Trinidad, wrote, the people, scarred with inferiority complexes, always felt they had to imitate outside cultures.

Remnants of European and United States colonial structures remain throughout the area. In fact, it is convenient to divide up and speak about the Caribbean along the lines of these colonial connections. For example, the Commonwealth Caribbean, the former British territories, consists of Barbados, Belize, Guyana, Jamaica, the Leeward islands (Montserrat, Antigua, St. Kitts-Nevis, Anguilla), Trinidad and Tobago, and the Windwards (Dominica, Grenada, St. Lucia, St. Vincent). Still other British-dominated islands are the Bahamas, Bermuda, the British Virgin Islands, the Caymans Islands, and Turks and Caicos. The French Caribbean includes the provinces of French Guiana, Martinique, and Guadeloupe. Half of the island of St. Martin is French-speaking, as is Haiti. The Spanish-speaking Caribbean is made up of the independent states of Cuba and the Dominican Republic, and the United States territory of Puerto Rico. The Netherlands Antilles include Aruba, Bonaire, Curaçao, Saba, St. Eustatius, Surinam, and half of St. Maarten, while the United States Caribbean consists of Puerto Rico and

251

the United States Virgin Islands. Other outside influences have also affected the region historically—e.g., the large slave populations brought from Africa and the East Indian communities, especially in Trinidad and Guyana.

The result is that Caribbean people speak European and Asian languages, as well as the local patois and Creole. This causes broadcasters enormous difficulties. On Netherlands Antilles islands such as Aruba and Curaçao, where the people are multilingual, speaking also the local Papiamento, mass media must perform in four languages. In other cases, broadcasters find it necessary to use patois, along with the language of the metropolitan country.

Political and economic systems in the Caribbean are also varied. Cuba is a Marxist country, as was Grenada before the United States invasion; the French Antilles remain as states of France; the former British colonies are parliamentary democracies; and Haiti and the Dominican Republic have been dictatorships in recent years.

The area has been accorded a great deal of attention in the past generation, with Castro's revolution; the United States embargo of Cuba and invasion of Grenada; the Bishop experiment in Grenada; the overthrow of Duvalier; revolts in Surinam and elsewhere; mass independence movements; the Caribbean Basin Initiative of Reagan; Radio Martí; the Jonestown massacre; and economic recessions in many countries, in one instance, leading to a nation virtually declaring bankruptcy.

BROADCASTING SYSTEMS IN THE CARIBBEAN

These myriad cultural influences are reflected in the broadcasting systems—their ownership patterns, technology, and programming use. For example, in the Commonwealth Caribbean alone, of twenty-seven radio stations, sixteen are owned by the state directly or through a public corporation; 5 are religious stations, owned by foreign evangelical groups; 5 are privately owned and 1 is a privately owned consortium (Cholmondeley, 1984, p. 15). The number of stations also shows a wide variance—for example, the Dominican Republic has 298, predominantly non-governmental; Puerto Rico, 62, all private; and Cuba, 57, all government-owned. Most other countries have one or two stations, except for Haiti (over 30), Netherlands Antilles (10), United States Virgin Islands (8), Barbados (4), and Surinam, Dominica, Bermuda, the Bahamas, and Antigua, 3 each.

Among countries with national television services, the Dominican Republic leads with 23; followed by Puerto Rico, 10; Netherlands Antilles, 4; and United States Virgin Islands, 3. Islands that have two services include Cuba, Bermuda, and Dominica; while Antigua, the Bahamas, Barbados, the British Virgin Islands, Grenada, Haiti, Jamaica, St. Kitts, St. Lucia, St. Vincent, Surinam, and Trinidad and Tobago each have one (see Wetzel, 1986).

For years, outside corporations set up, and then owned, many of the region's broadcast services; among these were Rediffusion of London, Thomson, and the United States broadcasting networks (for background, see Lent, 1977). Vestiges of foreign ownership remain, but as many islands became independent since the 1960s, so did their broadcasting services. Rediffusion retains some outlets, as do United States evangelical groups that have gospel stations on at least Anguilla, Antigua, St. Kitts, and Dominica. Some liberalization of licensing has taken place in the French Caribbean, as some private stations have been established with local ownership. But branches of French companies, such as Radio France Outremer or Télédiffusion de France, and relays of Radio France, still exist in French Guiana, Guadeloupe, and Martinique.

Especially since the mid-1980s, Puerto Rican television increasingly has become owned by mainland United States companies, partly because the island represents one of the bright spots for Latin American television, has relatively good TV production facilities, and offers the possibility of a site for a superstation to broadcast to the burgeoning Hispanic population on the mainland. The main station WAPA, for example, has been under the successive ownerships of Winston-Salem, Columbia Pictures, Western Broadcasting Company, and SFN Communications. Another major station, Telemundo, is owned by Harris of the United States while Malrite Communications, Lorimar, and James Leake, all of the United States, own other stations.

Technologically, Caribbean broadcasters heavily depend on outside corporations. Telecommunications infrastructures have been set up by North American and European multinational and transnational corporations, the major ones being Cable and Wireless, RCA Global Communications, TRT Telecommunications, ITT Communications, MCI International, AT&T, and Northern Telecom. Telecommunications equipment is imported from a number of countries, chiefly from the United States, Canada, Japan, and Great Britain, but also from France, Germany, Sweden, Holland, and Italy. At times, training schemes and programming have been part of the package that comes with the technology.

The dependence of Caribbean broadcasters upon foreign programming has intensified in the past decade. Despite some national and regional efforts to localize larger segments of broadcast schedules in the 1970s, the enticements that cable systems, satellite relays and overspills, and videocassettes provide in the way of inexpensive, easily accessible foreign programming are too overwhelming. In the Commonwealth Caribbean, the range of foreign show percentages vary from 50 to nearly 100 percent. Even countries such as Jamaica, which earlier had tried to stem the flow of foreign shows, now use over 88 percent non-Jamaican content. After Jamaican news and government information programs, the rest are foreign in origin. In Barbados, of the ten most popular television shows, all except the local nightly news are from the United States; Dominica's two television stations and one cable system depend almost solely on United States shows, as does Bahamian television.

The prevalence of ground stations linked to INTELSAT, and the Caribbean's location within the overspill of United States domestic satellites, enable local television services to pick up overseas signals easily. Besides individual home ownership of satellite dishes, entrepreneurs on most islands have set up relay stations to retransmit material directly from satellite through cable or a VHF station. On smaller islands that have national television services, these satellite-to-cable hookups have been threatening, to the extent that in Antigua and St. Kitts, the national service is given a space on the cable network. In other cases, where national television does not exist, such as Belize, Dominica, St. Lucia, Guyana, Turks and Caicos, Montserrat, and St. Vincent, the satellite-cable linkup dominates, using United States fare including commercials, news, and weather forecasts. On Montserrat, for example, a cable viewer might see CNN News, followed by United States situation comedies and detective shows, and news from WGN in Chicago, complete with advertisements for Chicago car dealers and the weather forecast for that city.

Most Caribbean territories now have cable television services. However, Puerto Rico is inundated with them; in 1986, at least 30 cable channels (24 in English) were available to more than 90,000 subscribers. Most receive their programming via satellite from the mainland; and the largest, Cable TV of Greater San Juan, is owned by Harris of the United States.

VIDEOCASSETTES IN THE CARIBBEAN

History

The influx of the videocassette recorders into the Caribbean came about without much fanfare; it just happened. In Jamaica, video became popular after the 1980 elections, when politicians campaigned with their own recorded videocassettes, which were done on an amateur level. Recognizing the potential of the VCR to bring in alternative programming to the sometimes-dull Jamaica Broadcasting Corporation schedule, and to record family events, a few individuals asked travelers to the United States to bring back sets. After a few video clubs were formed in Kingston, and then in Montego Bay and elsewhere, acting as viewing theaters and lending libraries, video technology seemed to catch on.

Video just seemed to appear elsewhere as well. In Belize, where people in border cities to Mexico and Guatemala had been pirating television since the mid-1970s, the home video became important when Belizeans living in the United States brought gifts of TVs and VCRs to relatives. Immediately afterward, the business community began to import them. According to a pioneer, Nestor Vasquez, owner of Tropical Vision and Channel 7, as well as chairman of the Belize Telecommunication Authority and member of Belize Broadcasting Authority, "a few industrious people had relatives in the United States tape shows such as "M.A.S.H.," which were brought in on a weekly basis by plane and rented to the public at Bz$5 (U.S.$2.50) each" (interview, Nestor Vasquez, Belize City, Belize, May 28, 1987). The original two rental shops—Nibble and Company, owned by Giovanni Smith, and another by Efrain Aguilar—were completely dependent upon relatives taping in the States and the airlines transporting videotapes in. Vasquez and a relative of his, American-born Emory King, decided that with an earth station, they could pick up the many United States stations and networks available in quick fashion, have 24-hour-a-day programming, and copy cassettes to rent to the public. King said:

> We realized too we could make money. We researched satellite companies and in October 1980, I went to Texas, brought back a satellite which I bought there. Our business went very, very well. We rented tapes at Bz$4 (U.S.$2) per night and in ten months, we were making Bz$10,000 gross income every month. There were enough VCRs in Belize to support all three of our rental shops (interview, Emory King, Belize City, Belize, May 29, 1987).

In 1981, Arthur Hoare also imported an earth station, as well as a transmitter; through his Coordinated Electronics, he began broadcasting live Mexican and United States programming. According to King, this "free" television had the effect upon videocassettes of an "elevator without power," plummeting to the bottom (interview, King). He added, "Vasquez said we too had to broadcast live, but I said I would not do it as it was illegal, and so I sold my shares in our company, Tropical Vision, to Vasquez" (interview, King).

During video's peak period in Belize in 1980–1982, as well as today with satellite-cable hookups, pirating has been rampant. However, unlike in most countries, pornography never found much of a market among the pirated works Belizeans chose. One dealer had a "few tapes under the counter, but pornography was not big here because people are sexually adjusted and you can't sell something that is free in a society," King said (interview).

Tropical Vision's two owners did not believe they were pirating by bringing in United States television via satellite. King explained:

> We don't think this is pirating. I thought it [satellite-relayed TV] was a good service of the United States, to give United States views to our region. I recall writing to a movie director friend after I set up the earth station, telling him I saw one of his movies here. He wrote back a blistering letter saying it was pirated. I was not aware it was unlawful. Actually, in 1980, I wrote a letter to every channel that our satellite was picking up—I remember in particular, at least HBO, WTBS, WGN, ESPN, Cinemax, and Showtime—and said, "Look, we bought this earth station, and we are renting tapes we make of your stuff. And we want to know what your royalty is." We got two responses to the over twenty letters sent out—from HBO and Cinemax. They responded, "We can't charge you anything. There is an international treaty to which the United States is a signatory, that allows each government to put up a satellite for its own domestic service and it cannot sell beyond its national borders. Therefore, we can't sell outside the national borders of the United States" (interview, King; see also Barry, 1984; Lapper, 1984).

King claimed that he was "amazed" when President Reagan's Caribbean Basin Initiative bill was passed with a "rider saying no country could get benefits of the CBI that used pirating" (interview, King).

As in most parts of the world, VCRs and cassettes started to become popular in the Caribbean at the beginning of the 1980s. Strange as it may seem, the two territories where home video had its most spectacular growth during that time were also among the most financially insecure—Puerto Rico and Guyana. Even with an economic recession and an unemployment rate of 25 percent, Puerto Rico had an explosion of video between 1980 and 1982 when a dozen video clubs were launched. In 1982 alone, they had U.S.$12 million in sales.

Guyana, which a few years ago was bankrupt, experienced a 14-fold increase in home video between 1980 and 1983. By 1983, there were nine video clubs with paid-up memberships of one hundred sixty, and an estimated 500,000 cassettes in circulation.

Penetration and Economics

Home video has been a relatively expensive medium in some countries. For example, Puerto Ricans pay a 19.8 percent import duty on video hardware; yet by 1984, about 150,000 to 200,000 VCRs and 250 video clubs were on the island. San Juan had 30 legal sales and rental outlets. Over 14.6 percent of the television homes, and 3.8 percent of the total population, possessed VCRs. Figures for other territories are equally high. Guyana, by 1983, had over 16,000 VCRs, representing 3.4 percent of the television homes; and Jamaica, by 1985–1986, 7.8 percent of television homes and 0.1 percent of the total population. The Dominican Republic's figures are relatively lower—2.5 percent of television households, and 0.2 percent of the population. One researcher blamed the poor state of the economy and the variety of the many television channels for the low penetration (Straubhaar, 1986, p. 14). Even among those who had the earning power, home video ownership was low.

In the beginning, Jamaicans became members of video clubs by purchasing tapes for J$140 to $170 (U.S.$77 to $93), for which they could borrow up to three tapes for J$50 (U.S.$27). Prerecorded tapes cost J$185 (US$102) for 1½-hour films and up to J$210 (U.S.$115) for foreign ones. Prices for VCRs have also been an exorbitant J$2,900 (U.S.$1,595) (see Thomas, 1983, p. 57).

Prices for home video products in Puerto Rico are rather high. A prerecorded motion picture tape is U.S.$75; a blank tape, $16; a VCR, $800 to $1,200; a giant 72-inch television system, $4,000. However, rentals of cassettes are decreasing in prices; most rent for $2 to $4 per day ("Puerto Rican Ban," 1982, p. 54). The Guyanese pay about U.S.$2,000 for a VCR and U.S.$60 for a prerecorded cassette.

According to a UNESCO estimate, the number of VCRs in Belize in 1980 was 1,883; but with the introduction of satellite-cable transmitted television, it went to only 2,500 by 1985. Still, only the United States, Canada, Panama, and Venezuela in the Western Hemisphere had more widespread videocassette viewing than Belize. The same survey showed there were 14,000 to 15,000 television receivers for a total population of less than 160,000 in Belize (in Petch 1987, pp. 13–14).

Belize is an important case study because it is a country where television was introduced solely via VCRs, and where today, at least nine on-air TV broadcasting stations and other cable services function, nearly 100 percent programmed by United States Stations. It is significant as well because these same satellite-cable linkups have destroyed the home video market. King said: "As soon as free TV came in, the VCR business died. It's all gone. I don't know what people use VCRs for now" (interview, Emory King, Belize City, Belize, May 29, 1987).

A survey of Belizean videocassette rental and electronics stores in mid-1987 found that only three small outlets rented prerecorded cassettes, and the market for VCRs was unstable. The largest rental shop is the pioneer Tropical Vision, which is "lucky to rent five a day," down over 90 percent from 1981–1982, when it did a business of at least 100 daily (interview, Nestor Vasquez, Belize City, Belize, May 28, 1987). When video was at its peak, Vasquez used six VCRs to record regularly from satellite. Today, at one of the rental shops, which is an adjunct to a furniture upholstery business, there is a stock of 75 to 100 cassettes, each renting for Bs$5 daily. Very few ever leave the shelves. Electronic store dealers reported varying interest in VCRs; one manager said he sold twelve in a month at about Bz$1050 (U.S.$525), while another sold only one.

Programming and Uses

It goes without saying that the primary purpose of the VCR in the Caribbean is to show imported television and film fare, mainly from the United States. As an example, in St. Lucia, television viewers can get twenty-four hours daily of United States programs through satellite downlinks and videotape rental stores. In Puerto Rico, where Spanish is the language of commerce and everyday conversation, the population seeks United States cassettes, even though they are in English without subtitles.

A number of reasons can be offered for the popularity of United States originated videocassettes. First, they are readily available, either in local outlets or through the large number of relatives and friends that West Indians have in the United States. Second, because of the nearness of the United States to the Caribbean, there has been a long tradition of cultural exchange and superimposition that dates back over 250 years. Third, United States television and film offer an alternative to some poorly produced local television that is often dull; the government-sponsored developmental content often focuses on family planning or nutrition. There are some exceptions in some territories; Trinidad's Banyan Experiment has produced interesting drama, as have Spanish stations in the Dominican Republic and Puerto Rico. Fourth, as is the case throughout the world, videocassettes are elusive, capable of being smuggled, pirated, and copied without much governmental control. In the Caribbean, most governments have not even attempted to regulate new technologies such as home video. A 1985 conference on communication and development called for the regulation of new technologies that bring in "massive penetration of our societies by foreign culture," but probably to no avail. Caribbean governments are reluctant to react strongly against outside cultural influences because they (1) expect the new information technologies to maintain social order for them, (2) are reluctant to remove a satisfying, status-giving activity for the influential classes, (3) give a low priority to communication policy in relation to other development projects, (4) have neither the time, skills, nor staffs to deal with control of media such as home video, (5) react to direct political pressures not to curb outside cultural fare, and (6) are hooked to the Caribbean Basin Initiative, which stresses more use of United States media contents (Caribbean Council of Churches, 1986, p. 5).

The very few surveys that include home video provide some not surprising findings. For example, in Puerto Rico, sex-oriented cassettes are popular, accounting for 20 to 30 percent of the rentals in 1982. This figure undoubtedly is higher after a 1983 ruling curbing sexually explicit content in cinema houses. Other popular videocassettes in Puerto Rico have been *Paternity*, *S.O.B.*, *For Your Eyes Only*, *Rocky*, as well as United States-produced classics such as *The Godfather* or *The Sound of Music*. One writer said that if a customer rented four videocassettes over a weekend, one would be pornographic, to be viewed with the wife; one Disney cartoon for the children; and two English-language films for the entire family ("Puerto Rican Ban," 1982, p. 54).

Seventy percent of Guyana's cassettes originate in the United States, while 15 percent come from Canada, 12 percent from the West Indies, 2 percent from Western Europe, and 1 percent from Venezuela and Brazil. A 1983 survey showed that Guyanese preferred romantic dramas, comedies, sports events, serials and musicals, horror shows, Westerns, courtroom dramas, and old movies (preferred by adults). Indian films were popular among the large East Indian population, and some pornographic cassettes were also popular in Guyana. Viewing times preferred for home video were: (a.) morning hours by housewives in suburban and middle-income urban areas; (b.) evening and night hours by professionals, public servants, and manual laborers; (c.) all day Saturday and Sunday by young people and children; and (d.) weekends (except Saturday morning) by working adults. Among uses viewers related were: source of entertainment, escape from reality, form of home education, and a means of uniting the multiethnic population and keeping children off streets (Forsythe, 1983).

In both Guyana and Puerto Rico, the safety of watching home video was cited for its popularity. A Guyanese wrote that home video was preferred to cinema attendance because of "less risk of one's vehicle being stolen outside the overcrowded cinemas." He also provided another reason for its popularity.

> And there is the notion, respected by Guyanese housewives, that video is there to keep women at home while, even in isolated cases *and* places, the roving husband enjoys *his* video shows in the second home of his so-called *essential* "deputy" wife (Forsythe, 1983, p. 53).

Other uses given for the VCR were its status-conferral characteristic, and its babysitting capacity, the latter mentioned in both Jamaica and Guyana.

Impact

The videocassette is blamed for the same impacts it supposedly has in other parts of the world: it deprives the government of taxation revenue and audiences for its developmental programming; it impinges upon local television and film industries; and it infringes upon traditional cultural and consumption habits.

A few Caribbean voices have begun to raise a hue concerning the potential cultural impact of the videocassette. One Caribbean Broadcasting Union official listed the "high and growing" penetration of VCRs as one of the six major dilemmas facing regional broadcasting (Rudder, 1986, p. 125). A Guyanese writer blamed VCRs for bringing into homes in his country, "irrelevant and questionable North American life-styles," which have an impact on

> clothes and hairstyles of the young, [it has] highlighted gangsterism; encouraged consumerism, through TV/video advertisements; and even contributed to the rate of immigration to North America (Forsythe, 1983, p. 53).

The complaint of United States-inspired consumerism brought in by television programs via satellites and videocassettes is often heard in Belize. One critic said that after viewing United States advertisements, Belizeans demanded imported United States products (Lapper, 1984, p. 16). Researching more specifically satellite-transmitted television in Belize, Oliveria found a relationship between television viewing and product preference. He said individuals who watched United States television tended to consume more United States products, while those who spent more time viewing Mexican television (also available in Belize) were more likely to consume regional products (Oliveria, 1986, p. 144). Because videocassettes are often made by copying from the television screen and therefore contain everything transmitted, including the commercials, Oliveria's points probably refer equally to them.

An impact that has not been studied very carefully is the role of media in migration. In Belize, where large numbers of people leave for the United States, a recent study showed that exposure to United States entertainment via television, and previously VCRs, was associated with a desire to emigrate among high socioeconomic status adolescents only. But news exposure was not significantly related to a desire to emigrate among young people (Roser, et al., 1986).

Belizeans attributed many impacts to VCRs and television. King said Belize's dress codes have been affected by videocassettes and TV penetration, explaining high school children's preferences for tuxedos and flowers at prom time, which was not the case before. He also believed the United States foreign policy is being sold to Belizeans through these media:

> Americans would be crazy to enforce the law against video and television being imported in. They are getting their point of view across to the Third World with a lot of Reagan and John Wayne and the like on TV. In fact, Congress would be wise to give United States taxpayer money to film companies as royalties to keep pumping this type of stuff to the world. We have always been pro-American here and the United States TV has certainly reinforced that. The impact is subtle—making one start to think like Americans. One leftist said TV doesn't just brainwash the people, but our government in Belize too. Leaders see Reagan doing things and they want to do them also (interview, Emory King, Belize City, Belize, May 29, 1987).

Listening to other Belizeans, one would gather that all aspects of life have been affected. Because WGN in Chicago is one of the stations taped or aired live, Belizeans are enthusiastic fans of the Chicago Cubs, plastering "WGN-Chicago Cubs" stickers on their automobile bumpers, writing or calling in their greetings to the Cubs' sportscaster, and attending games at which time they wave "Belizeans Love Cubs" banners. Other reactions have come in the form of complaints by irate husbands who said their lunches were unprepared or delayed as their wives watched noontime soap operas. In fact, soap operas were moved to a later time slot as a result.

The secretary of the Belize Broadcasting Authority, Agnes Ewing, saw an impact that video, and later TV, had upon socializing. She said that people are buying high-priced goods they cannot afford, that local products are taking a rear seat to foreign ones, and that fewer people are on the streets at night. Also, all movie theaters, except one, have closed (interview, Agnes Ewing, secretary, Belize Broadcasting Authority, Belize City, Belize, May 29, 1987). One other Belize City theater has been converted into a video palace, with a twenty-foot screen on which satellite-received movies and sporting events are shown.

An editor in Belize said she did not think video or television had negative effects upon newspaper reading, because only five minutes of local government propaganda appears on television, and Belizeans look to newspapers for information about their country (interview, Amalia Mai, editor, Belize Times, Belize City, Belize, May 29, 1987). However, the program director of Radio One, Belize's only radio service, said VCRs definitely affected that medium. He said:

> The VCR had a lot of impact upon radio listening, so much so that it caused us to change our programming at Radio One. When VCRs and television were novelties here, our listening plummeted. It is surging upwards now because we instituted a 60 percent Caribbean music rule and we air local shows most of the time. Those changes in our policy were conscious reactions to VCRs and TV (interview, Ed York, Belize City, Belize, May 29, 1987).

The one territory of the Caribbean where home video (in association with cable television and the economic recession) is blamed for damaging the film business is Puerto Rico. *Variety* ("Puerto Rican film attendance off," 1983, p. 37) reported that in 1983 alone, film billings decreased by 8 to 10 percent. The figure for 1981–1983 combined was a staggering 20 to 25 percent. Of course, a number of cinema houses have closed because of the drop in attendance.

SUMMARY

As in other parts of the world, the videocassette recorder has stealthily moved into the Caribbean, bringing with it outside television and film content (predominantly from the United States) with very little regard for international copyright regulations and domestic broadcasting and film businesses and their regulations. Perhaps the differences between the Caribbean and other Third World regions are that: (a.) the Caribbean already is so heavily saturated with new information technologies bringing in outside messages that the videocassette is hardly noticed and not accorded as much concern; and (b.) the Caribbean governments are even less capable of (and certainly less motivated to) regulating videocassettes.

PART 3

Conclusions and Future Prospects

CHAPTER 7

Conclusions and Future Prospects

Videocassette recorders have shown a rapidly growing popularity throughout most of the Third World, surpassing all other recent hardware innovations, except maybe radio and television. This is despite the obviously greater cost of a VCR (in addition to the television set itself) in poor countries, where cost is a major consideration. In fact, the popularity of VCRs is clearly related to that of television, but in different ways in varying situations.

In most places, a relatively expensive VCR is acquired because people, not happy with what is broadcast on television, are willing to part with scarce resources to "improve" or personalize their "television" viewing. In many places, television is seen as dull, inane, poorly produced, or propagandistic, particularly where broadcasting is government-controlled. In private systems, television may seem too commercial, too secular, or too frequently imported from abroad. For linguistic and ethnic minorities, broadcast television may simply be in the wrong language or represent the wrong religion, politics, or culture. For political activists, the VCR may open the door to alternative views not expressed on television news or public affairs programs. Most commonly in the Third World, the VCR represents an opportunity to get all the entertainment viewers want—whether movies, sports, music, or whatever—without the didactic programming offered by most development- or education-oriented government stations. Finally, in most Third World nations, possession of a VCR, even one held in common by a village or extended family, is a major status symbol, going beyond television itself, which long had been the status symbol.

Just as in the more industrialized nations of the Western world, people in the Third World prize VCRs because they enable them to individualize their viewing of video. This is threatening, though, to authorities who wish to use television to get across certain messages. It is equally problematic for communist parties, development planners, religious programmers, educators, government censors, and politicians seeking office or support. All planned television programming can be switched off in favor of whatever the VCR owner can rent, borrow, purchase, or record at an earlier time.

As for what people watch with VCRs, our interviews with video rental shop owners in a variety of Third World countries indicate that the most popular video choices are American feature films, other countries' popular films (such as those from India or Egypt), American television series, and other entertainment originally created for film or television in the major entertainment-producing nations. In a few places, people are beginning to add cameras to their VCRs to create their own video programs. Many of these videos, as a trend in Brazil, India, and Nigeria shows, are personal or family oriented: tapes of weddings, religious and community ceremonies, or sporting events. Some, however, are social and political (as Appendix 5.A indicates), produced by political, religious, labor, feminist, ecological, or neighborhood groups that do not have access to the mass media, but want to produce and distribute their own messages. Sometimes the content is intended to be local, as in the video bulletin boards in Nepal; other times the intent is to reach a national audience with alternative interpretations, as with the metalworkers' unions in Brazil.

A particularly troublesome but popular video type is pornography. Increased viewing of pornography has been singled out as a major problem of VCR use in Europe, the United States, Asia, Latin America, and the Arab nations. Estimates of the relative proportion of pornography among total video rentals or sales are varied and unreliable, but indicate that the privacy of video viewing at home, as well as the growth of specialized video parlors for pornography, have increased viewing of pornography among adults, adolescents, and even children. In several countries—for example, in the Philippines—pornography is both produced locally for the video market and imported from abroad.

With people tailoring their own video diet of entertainment, alternative political interpretations of events, and "immoral" sexual content, serious social and political implications have emerged. There is some evidence that children and adolescents are being exposed to more sex and violence via VCRs than on broadcast television, particularly in the many societies that control televised portrayals of sex and violence. If VCRs increase the overall time spent viewing television, then children and adolescents will also spend less time on reading, schoolwork, and other leisure pursuits.

Literary and cultural traditions are being affected by video.

Although VCRs may permit local productions to preserve dances, tales, or music, more evidence seems to indicate that they accelerate the basic effect of television in homogenizing culture on a national, regional, or international basis. Local culture, therefore, must compete with what comes through the screen. Social traditions also change. in some countries, such as those in Latin America, traditions of going out at night to stroll or go to bars or movies seem to have been changed by television and further changed by VCRs. In the Arab world, a tendency for most family members to stay home for entertainment has been reinforced. Throughout the world, people now party with a VCR, particularly since this confers status on the VCR owner in most Third World countries.

Video seems to affect different social classes differently. First, in most countries, VCR ownership is only possible for the elite. Second, because of this, VCRs are a trememdous consumer attraction for the middle class. In at least one country (the Dominican Republic) where we did extensive interviewing, more people owned VCRs than actually used them, since competition from cable TV and six broadcast television channels made the VCR largely unnecessary for entertainment; it was useful as a status symbol. Third, VCRs may be widening the gaps between social classes. In several Latin American countries and elsewhere, while the middle- and lower-class audiences are relatively happy with the nationally produced mass entertainment on broadcast television, elites are more likely to want imported American programs on video. Fourth, VCRs may thus be contributing to what dependency writers have termed the internationalization of the bourgeoisie—i.e., increasing the cultural, economic, and political ties of middle and upper classes not to their own countries, but to international cultural trends, multinational business, and foreign governments. Fifth, inasmuch as video programs have useful information in them, it may be that by having more access via VCRs, elites will gain more information to further widen existing gaps in information between classes, thus contributing to the retention of class boundaries.

It is already clear that video affects ethnic and linguistic minority groups. In many countries, such groups struggle to maintain traditional languages and cultures against pressures to assimilate into larger regional or national cultures and economies. Since television is frequently one of the major vehicles for promoting assimilation, VCRs are a tool for resisting assimilation. For example, in Malaysia, where television is used to promote Malay language and culture, the Chinese minority use video from Hong Kong, Taiwan, and other Chinese television-producing nations to maintain their language and culture. VCRs may also give various minorities—not only ethnic minorities, but women, activists, small political parties, and those with specialized or divergent interests—chances to acquire and produce specialized or alternative programs to fit their needs and, perhaps, even to get their views to other audiences. In the Arabian Gulf states, workers from India, Pakistan, and several Southeast Asian countries use VCRs to view native television programs and films while away from home.

Overall, it seems as though VCRs and video are a tremendous force for decentralizing and decontrolling "television" in the Third World. It may be that video helps level social differences by giving more people access to the production and distribution of "television." Conversely, it may be that video helps accentuate differences. In a way, both seem likely, since it seems certain that video has fractured the mass audience in many countries into various segments or groups. In some countries, however, the mass television audience has been sufficiently pleased by what is broadcast that the tendency toward acquiring VCRs has been limited to the elites.

How will video be controlled or governed? Some countries have simply tried to ban video and VCRs to avoid problems, but these attempts have failed, even in highly controlled societies such as the USSR. In setting up more realistic laws, the two principal concerns are censoring or controlling content and protecting copyrights against piracy for both national and foreign film and television industries. The Arab nations, the Philippines, the Chinese, and others have tried to set up strict controls on the video that enters their countries or is domestically produced for distribution. The main concern about content is with pornography, but general moral values, consumerism, and political issues also come up.

However pressing content concerns are in some countries, the main preoccupations with VCR uses and effects are actually economic. The main issue is the pirating of films onto video tape without paying film producers. This affects foreign film distributors who had seen the Third World as a lucrative market, now diminished considerably by video piracy. It also affects local or national film industries in a number of Third World countries such as Brazil, Egypt, Hong Kong, India, and Mexico, which produce for both national and foreign markets; they also lose major proportions of their revenue to domestic and international piracy. In India, the Philippines, Pakistan, and other Asian countries, the drop in film attendance attributable to video is very large. Furthermore, the once-profitable export market for Indian films has virtually collapsed.

It is clear that piracy of videocassettes lowers rental prices, adding to the diffusion of VCRs and the volume of rentals. An implication is that consumers (or viewers) will resist antipiracy moves that raise the price of rental cassettes. For example, the public reaction in Hong Kong was strong when a crackdown on piracy increased rental prices up to 500 percent. Furthermore, in China and the Arab nations, the concepts of copyright and its protection have not been particularly strong in traditional law; works of art are to be publicly shared and available. To some degree, videocassettes represent one of the most currently visible aspects of a larger debate on how to deal with trade in intellectual property between countries, and how to protect intellectual property rights within developing societies as they move increasingly into information-related industries.

Major questions arise concerning the uses of copyright laws. In Brazil, copyright laws protect the national film industry, but are vague on protection for foreign distributors. Governments are usually more concerned about protecting national, rather than foreign, industries. In Brazil and Venezuela, for instance, copyright changes are linked to channeling international distribution through national companies (usually broadcasters), who are made monopoly or oligopoly channels of legitimate distribution. This favors national industry but may result in internal media "imperialism" or, in other terms, extreme horizontal integration between film, video, and broadcast media.

The United States government and film industry, as well as national film and broadcasting personnel, have been pressuring for a crackdown on video piracy in a number of countries. United States pressures have included threats about bilateral trade sanctions and efforts to get films and other intellectual property issues included in multilateral agreements, like the General Agreement on Trade and Tariffs. Partially in response to this pressure, there have been attempts to update copyright laws to better protect films on video and foreign film/television works (notably underprotected in many countries), and to improve enforcement of existing laws in Singapore, Taiwan, and Venezuela, among others.

Even with improved copyright or intellectual property laws, enforcement remains inadequate because piracy is both easy and profitable, and the protection of copyrights and film master copies is intrinsically difficult. In some cases, special courts have been created, as in India, and more investigators and police have been put on the issue. Hong Kong, for instance, has fifty police on full-time piracy and copyright patrol.

Changes are being made in film/video commerce to facilitate enforcement. In several Asian countries, licensees offer sublicenses and attractive rates to pirate distributors for legitimate handling of films. Release times are being advanced, so that films are marketed simultaneously at home and abroad to reduce demand for quick, pirated copies of unreleased films. In some countries where film-going has come to be considered a lower-class activity, as in some Arab countries and South Asia, distributors are trying to make theaters more attractive. Governments and film industries are also attempting to anticipate audience interests in film and to put enough money into production so that revenues lost to piracy do not damage the quality of new films. Governments themselves, in many places, have had to confront relatively major losses in revenue from import duties and the taxes on cinema admissions and exhibition.

There are other economic issues created by VCRs, particularly for television broadcasters and their sponsors, governments, and advertisers. As broadcasters lose part, or even most, of their audience to video, as seems to be happening in Taiwan and the Arab Gulf states, for instance, the implications are economic as much as political. Advertisers may cut back investment in television, and governments may wonder about the efficacy of funds put into the medium, although the tendency in most countries, such as the Gulf States, Malaysia, Singapore, or Thailand, has been to expand both program production and the number and diversity of channels to try to win back audiences. Fiji rushed into a national television system because of video encroachments on the potential audience; Venezuela and Colombia brought in color television after people demanded it because they were used to seeing videocassettes in color. These kinds of investments divert money from other worthy government projects.

Advertisers are beginning to put ads directly into videocassettes. The process is crude in Egypt and elsewhere, but will become more sophisticated. In several countries, including Brazil, commercial product exposure is now being built into film and television scripts and visuals. As this process becomes more widespread, it may draw revenue away from conventional television advertising. This will also have an inpact upon countries that try to control advertising. Malaysia tries to restrict alcohol, pork, and baby formula advertisements, while Indonesia has banished all advertisements in an effort to reduce consumerism. What will happen to such controls if advertising-laden imported videos bypass them?

Television in many Third World countries is clearly reacting to video. First, at a structural level, a number of broadcasters, particularly in Latin America where most are private companies, are acquiring or integrating themselves with film and video distribution companies. They do this partially to market their own productions on video and to get into what seems to be a profitable new business. Broadcasters also do this, as in Bangladesh, Brazil, and the Philippines, by getting directly into the production of original videocassettes. Second, television networks are trying to offer programming that video is not noted for on the theory that "good television drives out bad video." The types of programming on television perceived to have an advantage vary: in Taiwan, children's programs, sports, local soap operas, and first-run foreign films; in Brazil, soap operas, sports, news, and interviews; in Saudi Arabia, news, sports, and children's programs. There is, however, an increasing amount of South-South interaction in television's fight with video. India sends experts to Mexico to learn how to make soap operas with a partially hidden educational content; small Latin American nations buy Spanish-language soap operas from larger Latin nations; the Arab Gulf states cooperatively fund and produce a popular adaptation of "*Sesame Street.*" These moves have interesting implications. For example, will the United States hegemony over television traffic diminish even while the United States presence in videos of films increases? Will there still be a homogenization of programming, with similar-looking soap operas produced locally? What happens to popular culture forms when they are exported, such as the Mexican *telenovela* to India?

Third, television seems to be reacting to video by privatizing, although other forces, such as the increasing globalization of consumer markets, also push privatization. Private channels are usually more entertainment oriented, and the Latin American experience shows that more entertainment on broadcast television tends to decrease the attraction of video and VCRs. With or without privatization, the number of television channels is being increased to confront video with more diverse programming. Both Malaysia and Saudi Arabia have opened new channels to compete with video, featuring more United States programs and, in the case of Malaysia, more Chinese-language programs for that minority. A question raised is, will this imply an increased inflow of United States programming, with more homogenization and cultural imperialism?

Fourth, as television channels increase, they may segment audiences into smaller, more specifically targeted groups. Mexico and Brazil have done this, separating children and adolescents, middle-class adults, and the broad general audience into distinct groups, or separating the upper, middle, and lower classes. If more television better fits specific interests, then the role for video may decrease. This process of preempting VCRs is, or will be, even more notable in those countries where satellite dishes and satellite-fed cable TV have increased diversity and segmented the audience even further, as in Belize or the Dominican Republic.

Fifth, the diversification and nationalization of television content seems to reduce use of video. The quantity and quality of national television seems to affect VCR diffusion, as in Brazil, where VCR penetration is much lower than in economically comparable countries. Sixth, television frequently can be made more interesting—i.e., more entertaining. At least in Latin America, it seems that more entertainment-oriented systems produce lower rates of VCR diffusion. All of this leads to yet another question: What happens to educational and informational goals predominant among Third World broadcasters? One solution is what India has done—sent teams of writers and producers to Mexico to learn how to produce soap operas that simultaneously entertain and teach lessons about family planning or other educational topics.

A last economic consequence of video concerns its hardware. Planners in a number of countries tend to see money spent on VCRs and videocassettes as money wasted on luxuries. However, attempts to restrict or impose substantial tariffs on video equipment have resulted in a large international smuggling and black market operations, as in the cases of Mexico and Brazil. Still it is clear that this informal, but substantial, investment in video by millions of owners and middlemen draws money away from other investments and uses.

What are the major trends in video in the Third World that might continue? It seems clear that VCRs will be used by ethnic minorities to keep in touch with their mother cultures, as the Koreans in the United States, the Japanese in Brazil, and the Chinese in Malaysia are already doing. Video will continue to play a role in introducing television into remote and rural areas, and even into small or poor countries, such as Fiji or Tanzania. Particularly for those countries that had tried to avoid the expense of a television system, this will come at some cost.

Television will react to video. Other cultural industries will grow to feed an increasing need for more national television production, as has happened in Brazil, where television has become a major support of the film, music, and theater industries. More video industries, legal and illegal, will be created, although some television industries may rush to fill or reoccupy this space. Video will continue to emphasize those things banned by governments or underplayed on television, whether it is news, as in Chile, or entertainment, as in Saudi Arabia. This has forced some governments to loosen up on both news and entertainment, and will probably continue to pressure others. Governments will also continue—or in some cases, begin—to crack down on video piracy. It is certain, though, that both legal and illegal video operators will continue to organize, as they have in Malaysia and the Philippines, to fight against national government restrictions and pressures from foreign distributors. Some video entrepreneurs defend piracy of Western films and television programs on the grounds that the West took their culture and artifacts and that what they are doing is justified.

Audiences will be affected further as the amount of time spent in front of the television set continues to grow. Although video viewing frequently substitutes for some television viewing, it also tends to increase the total time spent with "television." Other countries may come to have patterns similar to the seven-hour television days now spent by the average United States citizen. This will have an impact on work, leisure, and values. In particular, consumer expectations are likely to continue to rise, as viewers are exposed to more of the life-styles found in more affluent countries, and as advertisements or product exposures increasingly appear in the films and programs themselves.

As with other media, though, audiences are not totally passive in front of the television set. Much of video's impact will be related to existing behavior patterns and values. If people in a culture prize going out at night, as in Hong Kong (where apartments are tiny), or Latin America (where walking and socializing is an ingrained tradition), then staying at home with a VCR may not be attractive. A great deal also depends on what people feel to be lacking in their media consumption or in their lives more generally. If people are reasonably well informed and entertained (by their own standards), then the "marginal utility" of a relatively expensive VCR (by Third World income standards) may be too low to attract people to video. This seems to have been the case in several Latin American countries.

The ultimate effect of the home videocassette recorder on the developing world will not be known for some time. However, as we have discussed in the preceding pages, the videocassette phenomenon has spread rapidly in both richer and poorer sections of the Third World, and already the effect has been profound.

References

Aamoudi, K. A. al: (1984). *Toward the implementation of Saudi Arabia's information policy*. Unpublished master's thesis, Michigan State University, East Lansing, MI.

Abboud, M. (1986, March/April). Lifestyle boosts video popularity. *TV World*, p. 31.

Adwan, N. (1985, April). *Research on video programs in Iraq, Kuwait, and Qatar*. Baghdad, Iraq: Arab Center for Audience Research.

Agrawal, B. (1983). The socio-political implications of communication technologies in India. *Nedia Development, 3*, 26–28.

Agrawal, B. (1984). *Technological vector of video cassette recorders in the cultural context of India: Some observations*. Paper presented at meeting of International Association for Mass Communication Research, Prague, Czechoslovakia, August 27–September 1.

Agrawal, B. (1986). Cultural response to communication revolution: Many modes of video use in India. *Gazette, 38*, 29–41.

Alisky, M. (1983). *Latin American media: Guidance and censorship*. Ames, IA: Iowa State University Press.

AMCB. (1987, March–April), p. 4.

Anderson, C. E. (1979, May). The profession of videotape recording. *Broadcast Engineering*, pp. 60–64.

Angus, R. (1973, March). 75 years of magnetic recording. *High Fidelity Magazine*, pp. 42–45.

Antola, L., & Rogers, E. (1984). Television flows in Latin America. *Communication Research, 11* (2), 183–202.

Anuar, M. K. (1986, April-June). The Kampung Memali video "revisited." *Malaysian Journalism Review*, pp. 11–12.

Arabic broadcasts from London. (1938, January). *Great Britain and the East, 13*.

Atkin, C. K. (1985). Informational utility and selective exposure to entertainment media. In D. Zillman and J. Bryant (Eds.), *Selective exposure to communication* (pp. 63–91). Hillsdale, NJ: Erlbaum.

Baboulin, J. C., Gaudin, J. P., & Mallein, P. (1983). *Le magnétoscope au quotidien: Un demi-pouce de liberté.* Paris: Auber Montaigne.

Bakhaider, B. (1981). *The impact of the video cassette recorder on Saudi Arabian television and society.* Unpublished master's thesis, San Diego State University, San Diego, CA.

Bakr, Y. A., Labib, S., & Kandil, H. (1985). *Development of communication in the Arab states: Needs and priorities.* Paris: UNESCO.

Banerjee, S. (n.d.). *The role of videotape in rural development.* Unpublished document, Asian Mass Communication Research and Information Center, Singapore.

Barlow, G., & Hill, A. (Eds.). (1985). *Video violence and children.* London: Hodder and Staughton.

Barry, J. (1984, March). The Belize dilemma. *Media in Education and Development,* pp. 11–13.

Basu, K. (1984, September 16). Video pirates are down but not out. *Sunday Observer* (India), p. 40.

Belbase, S. (1985). *Uses of video for development in Nepal.* Paper presented at DEF International, Women in development video exploration workshop, Nairobi, Kenya, July 9–20.

Belbase, S. (1986). *Rural women in participatory communication* (Workshop 1). Kathmandu: Worldview International Foundation, Nepal, October.

Beltran, L. R. (1978). TV etchings on the minds of Latin Americans. *Gazette, 24,* 61–85.

Benoist-Mechin, J. (1958). *Arabian destiny.* Fairlawn, NJ: Essential Books.

Berrigan, F. J. (1979). *Community communications: The role of community media in development* (Reports and papers on mass communication No. 90). Paris: UNESCO.

Bibliowicz, A. (1982, June 24). La necessidad de las opciónes multiples. *Tele Revista* (Bogotá).

Birks, J. S., & Sinclair, C. A. (1979). Migration and development: The changing perspective of the poor Arab countries. *Journal of International Affairs, 33*(2), 285–308.

Blumler, J. G., & Katz, E. (Eds.). 1974. *The uses of mass communications: Current perspectives on gratifications research.* Beverly Hills, CA: Sage.

Boyd, D. A. (1970/71). Saudi Arabian television. *Journal of Broadcasting, 15,* 73–78.

Boyd, D. A. (1972). *An historical and descriptive analysis of the evolution and development of Saudi Arabian television: 1963–1972.* Unpublished doctoral dissertation, University of Minnesota, Minneapolis, MN.

Boyd, D. A. (1975). Development of Egypt's radio: "Voice of the Arabs" under Nasser. *Journalism Quarterly, 52,* 645–653.

Boyd, D. A. (1982). *Broadcasting in the Arab world: A survey of radio and television in the Middle East.* Philadelphia, PA: Temple University Press.

Boyd, D. A. (1985). The Janus effect? Imported television entertainment programming in developing countries. *Critical Studies in Mass Communication*, 1(4), 379–391.

Boyd, D. A. (1985). VCRs in developing countries: An Arab case study. *Media Development*, 32(1), 5–7.

Boyd, D. A., & Straubhaar, J. D. (1985). Developmental impact of the home video cassette recorder on Third World countries. *Journal of Broadcasting and Electronic Media*, 29(1), 5–21.

Boyd-Barrett, O. (1980). *The international news agencies*. Beverly Hills, CA: Sage.

Brito, B. (1985). La video comercialización en Venezuela. *Comunicación*, 49–50, 53–57.

Broadcasting and television law. (1976). Promulgated by the President on January 8, 1976, Taipei.

Calderon, E. L. (1986, August 24). Up in arms against video piracy. *Sunday Times Magazine* (Manila), p. 20.

Camargo, N. de (1975). *Communication policy in Brazil*. Paris: UNESCO.

Caribbean Council of Churches. (1986). *Draft reports and proposals: CCC-Inter media cooperation on communication for human development*. Unpublished.

Cassette scandal. (1980, May 16). *Asiaweek*, p. 21.

Censors to check video tapes. (1979, December 22–28). *Gulf Mirror*, p. 5.

Cheung, B. (1984). Broadcast piracy in Hong Kong. *Media Asia*, 11(1), 17–19.

Chiruvolu, P. (1986). *Mass media and participatory development: A study of Kheda*. Unpublished master's thesis, Michigan State University, East Lansing, MI.

Cholmondeley, H. (1984, January–March). Communications in the Caribbean. *COMBROAD*, pp. 12–16.

Choucri, N. (1986). Asians in the Arab world: Labor migration and public policy. *Middle Eastern Studies*, 22(2), 252–272.

Clad, J. (1985, February 28). The multitudinous media. *Far Eastern Economic Review*, pp. 21–23.

Copyright law of the Republic of China. (1986). Taipei: Ministry of Interior.

Datafolha staff. (1986, January 20–July 29). Filmes mais retiradas em video-clubes e locadoras. *Folha de São Paulo* (São Paulo).

Democracy in Communication. (1986). *Popular video and film in Latin America. (Guide to) a traveling exhibition of independent video and film, produced by Latin Americans*. New York: Ragged Edge Press.

DoBrow, J. A. (1986). *The social and cultural implications of the VCR: How VCR use concentrates and diversifies viewing*. Unpublished doctoral dissertation, University of Pennsylvania, Philadelphia, PA.

Dorfman, A., & Mattelart, A. (1975). *How to read Donald Duck: Imperialist ideology in the Disney comic*. New York: International General.

Eddy, W. (1963). King ibn-Sa'ud: "Our faith and your iron." *Middle East Journal*, 17, 257–263.

Einsiedel, E. (1986). *Video use in a small Philippine urban municipality: A case study*. Paper presented at International Association for Mass Communication Research meeting, New Delhi, India, August 25–29.

Enforcement rules of the broadcasting and television law. (1983). Taipei: Government Information Office, August.

Extent of piracy of sound recordings worldwide in 1984. (1985). London: International Federation of Phonogram and Videogram Producers.

Fejes, F. (1980). The growth of multinational advertising agencies in Latin America. *Journal of Communication, 30*(4), 36–49.

Ferreira, L., & Duke, D. (forthcoming). Broadcasting in Colombia. In E. Mahan & J. D. Straubhaar (Eds.), *Broadcasting in Latin America*.

Festa, R., & Santoro, L. (1987). Policies from below—Alternative video in Brazil. *Media Development, 34*(1), 27–30.

Forsythe, V. L. C. (1983). Guyana: Video parties—and fewer cars stolen. *Intermedia, 11*, 52–53.

Fox de Cardona, E. (1975). Multinational television. *Journal of Communication, 25*(2), 122–127.

Frey, F. (1973). Communication and Development. In I. de Sola Pool, F. Frey, W. Schramm, N. Maccoby, & E. Parker (Eds.), *Handbook of Communication*. Chicago: Rand McNally.

Gandelman, H. (1985). *The piracy of cinematic films on videocassettes*. Rio de Janeiro, Brazil: União Brasileira de Video.

Ganley, G., & Ganley, O. (1986). *The political implications of the global spread of videocassette recorders and videocassette programming*. Cambridge, MA: Program on Information Resources Policy, Harvard University.

Giron, M. (1986, October 22). Filipino video piracy virtually 100%. *Variety*, p. 443.

Gopal, B. V. (1986). *Video piracy and the law*. New Delhi: Deep and Deep Publications.

Greenberg, B. G., & Heeter, C. J. (1987). VCRs and young people. *American Behavioral Scientist, 30*(5).

Guback, T., Varis, T., et al. (1984). *Transnational communication and cultural industries* (Reports and papers on mass communication No. 92). Paris: UNESCO.

Gubern, R. (1985). La antropotronica: Nuevos modelos tecnoculturales de la sociedad mass-mediatica. In R. Rispa (Ed.), *Nuevas tecnologías en la vida cultural Española*. Madrid: FUNDESCO.

Guerra al Betamax. (1984, November). *Semana* (Bogotá).

Gulf crackdown on porn. (1986, May). *TV World*, p. 110.

Halleck, D. (1984, November). Notes on Nicaraguan video libre o morir. *The Independent*, pp. 12–17.

Hamid, F. (1987, July). Thai TV—Portrait of a medium. *Asian Advertising and Marketing*, pp. 28–33.

Head, S. W. (1985). *World broadcasting systems*. Belmont, CA: Wadsworth.

Heeter, C. (1984). *Explaining the video boom: Factors related to national VCR penetration*. Unpublished manuscript, Michigan State University, Department of Telecommunication, East Lansing.

History of the U.A.R. radio since its establishment in 1934 until now. (1970, August). *Arab Broadcasts*. Translated by Fatima Barrada.

Holloway, N. (1986, April 24). Pirates repelled. *Far Eastern Economic Review*, pp. 58–60.

Homevid is blamed for tax drop. (1981, December 16). *Variety*, p. 26.

Huang, Hamilton. (1986, August 4). *Free China Journal*, p. 3.

IBOPE/Instituto Brasileiro de Opinião Publico e Estatistica. (1984). *Levantamento socioeconomico de cidades* (Socioeconomic baseline survey for major cities). Rio de Janeiro: IBOPE.

Industria de lo audiovisual. (1985). *Comunicación, 49–50*, 14.

International Association for Mass Communication Research (IAMCR). (1985). *Big sounds for small peoples* (videocassette). Author.

Izcaray, F. (1984). *Some remarks on the agenda for communication research in the 80's: The impact of the new information technologies on Latin American societies*. Paper presented at International Association for Mass Communication Research meeting, Paris, August.

Izcaray, F. (1985, May). *Market survey of Barquisimeto*. Unpublished survey data. Barquisimeto, Venezuela.

Jabbar, J. (1983). A cautious welcome. *Intermedia, 11*, 65–66.

Jamison, D., & McAnany, E. (1976). *Radio for education and development*. Beverly Hills, CA: Sage.

Johnson, R. (1986, October). Film in Pos-Abertura Brazil. In panel on Pos-Abertura Cultural Policy in Brazil, Latin American Studies Association meeting, Boston.

Johnson, R., & Stam, R. (1982). *Brazilian cinema*. Rutherford, NJ: Fairleigh Dickinson University Press.

Jouhy, E. (1985). New media in the Third World. In E. Noam (Ed.), *Video media competition* (pp. 416–439). New York: Columbia University Press.

Kamaluddin, S. (1982, February 5). The blue movie blues. *Far Eastern Economic Review*, p. 98.

Katz, E., & Wedell, G. (1977). *Broadcasting in the Third World*. Cambridge, MA: Harvard University Press.

Kaye, L. (1984, May 24). Copycats unrepentant. *Far Eastern Economic Review*, pp. 82–83.

KFSH TV Guide. (1979, March 10–12). Riyadh, Saudi Arabia.

Koh, F. (1982, September 24). A dilemma in view. *Far Eastern Economic Review*, pp. 46–48.

Koppes, C. R. (1976). Captain Mahan, General Gordon, and the origins of the term Middle East. *Middle Eastern Studies, 12*, 95–98.

Kurian, G. (1984). *World Press Encyclopedia.* New York: Facts on File.

Kurukulasuriya, L. (1983). The rich relax with imported video. *Intermedia, 11,* 68–69.

Lapper, R. (1984, November). From pirate to private—with government approval. *TV World,* pp. 16–17.

Lee, C. C. (1980). *Media imperialism reconsidered: The homogenizing of television culture.* Beverly Hills, CA: Sage.

"Legal" homevid in Latino orbit. (1985, March 20). *Variety.*

Lent, J. A. (1977). *Third World mass media and their search for modernity: The case of Commonwealth Caribbean, 1717–1976.* Lewisburg, PA: Bucknell University Press.

Lent, J. A. (1978). *Broadcasting in Asia and the Pacific.* Hong Kong: Heinemann Educational Books.

Lent, J. A. (1982, October). How broadcasting operates in the ASEAN countries. *Index on Censorship,* pp. 6–9.

Lent, J. A. (1984). A revolt against a revolution: The fight against illegal video. *Media Asia, 11*(1), 25–30.

Lent, J. A. (1985). Video in Asia: Frivolity, frustration, futility. *Media Development, 32*(1), 8–10.

Leo, C. (1986, November 17). Making up ground against the "pirates." *New Straits Times* (Kuala Lumpur), p. 15.

Lerner, D. (1958). *The passing of traditional society.* New York: The Free Press.

Lhoest, H. (1983). *The interdependence of the media* (Council of Europe mass media files No. 4). Strasbourg, France: Council of Europe.

Li Bo. (1983). Gifts from overseas. *Intermedia, 11,* 44.

Lins da Silva, C. E. (1985). *Muito alem do jardim botanico.* São Paulo: Summus.

Litewski, C. (1985). Television, the IMF, and Brazil: An interview with TV Globo executives. In *TV & Video International Guide* (pp. 27–30). New York: Zoetrope.

Litewsky, C. (1985). Brazil. In *TV & Video International Guide* (pp. 45–48). New York: Zoetrope.

Lubis, M. (1983). A foreign attraction. *Intermedia, 11,* 54.

McAnany, E. G. (1987). Cultural policy and television: Chile as a case. *Studies in Latin American Popular Culture, 6.*

McBride Commission. (1980). *One world, many voices.* Paris: UNESCO.

McLuhan, M., & Fiore, Q. (1967). *The medium is the message.* New York: Bantam Books.

McPhail, J. C. (1981). *Electronic colonialism: The future of international broadcasting and communication.* Beverly Hills, CA: Sage.

Mahan, E. (1984). Government-industry cooperation and conflict in Mexico and the United States: A comparative analysis of commercial broadcast regulation. *Studies in Latin American Popular Culture, 3,* 1–16.

Mahan, E. (1985). Mexican broadcasting: Reassessing the industry-state relationship. *Journal of Communication*, *35*(1), 60–75.

Mahan, E., & Straubhaar, J. D. (forthcoming). The Dominican Republic. In E. Mahan and J. D. Straubhaar (Eds.), *Broadcasting in Latin America*.

Manning, R. (1984, May 24). Pirates under fire. *Far Eastern Economic Review*, pp. 62–63.

Mattelart, A., & Schmucler, H. (1985). *The new communication technologies: Freedom of choice for Latin America*. Norwood, NJ: Ablex.

Mattos, S. (1984). Advertising and government influence on Brazilian television. *Communication Research*, *11*(2), 203–220.

Mayo, J., et al. (1987, February). *Peru Rural Communication Services Project—Final evaluation report*. Center for International Studies, Learning Systems Institute, Florida State University.

Menon, U. K. (1984, December). The film industry is losing its grip. *Sasaran* (Malaysia), pp. 30–33.

Middle East Advisory Group. (1983, March). *Finding of MEAG survey*. Cairo, Egypt.

Middle Eastern story-tellers give way to videos. (1985, February) *Index on Censorship*, p. 53.

Mideast entertainment void prompts homevideo bonanza. (1981, May 13). *Variety*, p. 370.

Miller, J. (1984, February 2). Upset by "Sadat," Egypt bars Columbia films. *New York Times*, p. 1.

Mody, B. (1979). Programming for S.I.T.E. *Journal of Communication*, *29*(4), 90–98.

Mody, B. (1986). The receiver as sender: Formative evaluation in Jamaican radio. *Gazette*, *38*, 147–160.

MPEAA hits Caracas pirates on copyright. (1983, December 21). *Variety*.

Mullin, J. (1979, May). Discovering magnetic tape. *Broadcast Engineering*, pp. 80, 82.

Mustafa, M. N. (1987, March). Music videos for sale by Bangladesh Radio-TV. *World Broadcast News*, p. 18.

Murder of King Faisal. (1975, April 7). *Newsweek*, pp. 21–23.

Narwekar, S. (1985, July 21). Who's afraid of the video-monster? *Indian Express*, p. 40.

Ninan, T. N., & Singh, C. U. (1983, September). India's entertainment revolution. *World Press Review*, pp. 58–59.

Noam, E. N. (Ed.). (1985). *Video media competition*. New York: Columbia University Press.

No hiding place. (1986, April 24). *Far Eastern Economic Review*, pp. 58–59.

NTC/NCT Newsletter—Nuevas tecnologías de comunicación/New communication technologies. (1986). *1*(1). Instituto para América Latina (Lima).

Ogan, C. L. (1985). Media diversity and communication policy: Impact of VCRs and satellite TV. *Telecommunication Policy, 9*(1), 63–73.

Olivieri, A. (1985). Una nueva televisión está por nacer. *Comunicación, 49–50,* 25–31.

Oliveira, O. S. (1986). Satellite TV and dependency: An empirical approach. *Gazette, 38,* 127–145.

Oofy, A. al: (1986). *The impact of the video cassette recorder on young Saudi television-viewing habits and life style.* Unpublished master's thesis, Michigan State University, East Lansing, MI.

O'Sullivan, J. (1985). Community video fights for Latin American culture. *Media Development, 1,* 24–26.

Pacheco, E. B. (1986, August 21–27). Video business: The birth pangs of legitimization. *Veritas* (Manila), p. 22.

Palestine Department of Posts and Telegraphs annual report. (1936). Jerusalem.

Pan Arab Research Center. (1982, October). *Kuwait media survey.* Kuwait: PARC.

Pan Arab Research Center. (1982, November). *Basic media survey, Saudi Arabia.* Kuwait: PARC.

Pan Arab Research Center. (1986, Spring). *Bahrain media index, video section.* Kuwait: PARC.

Pasquali, A. (1976). *Comunicación y cultura de masas.* Caracas: Monte Avila.

Petch, T. (1987). Television and video ownership in Belize. *Belizean Studies, 15*(1), 12–14.

Pipes, D. (1983). *In the path of God.* New York: Basic Books.

Pool, I. de S. (1961). Mass media and politics in the modernization process. In L. W. Pye (Ed.), *Communications and political development* (pp. 234–253). Princeton: Princeton University Press.

Puerto Rican ban: theatre goers defect to homevid clubs. (1982, May 19). *Variety,* P. 54.

Puerto Rican film attendance off: blame table, homevideo growth, plus economy, rising crime rate. (1983, March 30). *Variety,* p. 37.

Quebral, N. (1985, July). How will video recorders be used in the villages? *Action,* pp. 4–5.

Ramada News. (1982, January). Doha, Qatar.

Rao, L. (1983). An Indian perspective on modern technologies of communication. *Media Development, 3,* 36–37.

Rao, L. (1986). *VCR and its influence on life styles in South India.* Paper presented at AMIC-CSD-WACC consultation, New Delhi, August 21–23.

Read, W. H. (1976). *America's mass media merchants.* Baltimore: Johns Hopkins University Press.

Redmont, B. S. (1985). Inside Chinese television: A new "giant leap forward." *Television Quarterly, 21*(3).

Robinson, D. (1985, December 4). Revisionist Rambo. *The Times* (London), p. 6.

Rodriguez, N. (1986, August 23). Sazon's package of 5 solutions to film piracy. *The Tribune* (Manila), p. 11.

Roe, K. (1983). *The influence of video technology in adolescence* (Media panel report No. 27). Lund: University of Lund, Department of Sociology.

Rogers, E. M. (1976). *Communication and change: A critical review.* Beverly Hills, CA: Sage.

Rogers, E. M. (1978). The rise and fall of the dominant paradigm. *Journal of Communication, 28*(1), 4–69.

Rogers, E. M. (1983). *Diffusion of innovations* (3rd ed.). New York: The Free Press.

Rolo, C. J. (1941). *Radio goes to war.* New York: G. P. Putnam's Sons.

Rosenblum, M. (1979). *Coups and earthquakes: Reporting the world for America.* New York: Harper.

Rosengren, K. E., Wenner, L. A., & Palmgreen, P. (1985). *Media gratifications research: Current perspectives.* Beverly Hills, CA: Sage.

Roser, C., Snyder, L. B., & Chaffee, S. H. (1986). Belize release me, let me go: The impact of United States mass media on emigration in Belize. *Belizean Studies, 14*(3), 1–30.

Rota, J. (1985). The content of Mexican commercial television: 1953–1976. *Studies in Latin American Popular Culture, 4.*

Rothnie, D. (1987, July). Black market thrives in a golden land. *Asian Advertising and Marketing,* pp. 46–47.

Rudder, M. (1986). Broadcasting in the Caribbean: A unique experience. *Third Channel, 1*(1), 125–135.

Santoro, L. F. (1985). *Video in Brazil.* Unpublished paper, University of São Paulo.

Sarathy, M. A. P. (1983). Video on the bus. *Intermedia, 11,* 53.

Scharfenberg, E. (1985, June). Piratas del siglo XX—Video. *Fama* (Caracas), 55–56.

Schiffman, S. (1984, March). Video piracy's real winners and losers. *TV World,* p. 32.

Schiller, H. I. (1969). *Mass communication and American empire.* New York: Augustus Kelley.

Schnitman, J. A. (1984). *Film industries in Latin America.* Norwood, NJ: Ablex.

Schramm, W. (1964). *Mass media and national development.* Stanford, CA: Stanford University Press.

Shobaili, A. S. (1971). *A historical and analytical study of broadcasting and press in Saudi Arabia.* Unpublished doctoral dissertation, Ohio State University, Columbus, OH.

Siebert, F. S., Peterson, T., & Schramm, W. (1956). *Four theories of the press.* Urbana, IL: University of Illinois Press.

Siguion-Reyna, A. (1986, August 21–27). VRB wake up! *Veritas* (Manila), p. 30.

Silverman, M. (1986, March 5). Mainland China tries video distribution. *Variety*, p. 1.

Smale, A. (1983, April 10). Soviets battle black market in western movie cassettes. *Philadelphia Inquirer*, p. 5–I.

Sodré, M. (1977). *O monopólio da fala*. Petrópolis, Brazil: Vozes.

Stangelar, F. (1985). Comunicación alternativa y video-cassette: Perspectivas en América Latina. *Comunicación, 49–50*, 58–70.

Straubhaar, J. D. (1981). *The transformation of cultural dependence: The decline of American influence on the Brazilian television industry*. Unpublished doctoral dissertation, Fletcher School of Law and Diplomacy, Tufts University.

Straubhaar, J. D. (1984). Brazilian television: The decline of American influence. *Communication Research, 11*(4), 221–240.

Straubhaar, J. D. (1985). *Broadcasting as a cultural industry in Latin America*. Paper presented at Latin American Studies Association conference, Albuquerque, New Mexico, April.

Straubhaar, J. D. (1986). *The impact of videocassette recorders on broadcasting in Brazil, Colombia, Dominican Republic, and Venezuela*. Paper presented at Studies in Latin American Popular Culture conference, New Orleans, April 10–12.

Straubhaar, J. D. (forthcoming). TV flows among Latin American countries. In E. Mahan and J. D. Straubhaar (Eds.), *Broadcasting in Latin America*.

Straubhaar, J. D., & Lin, C. R. (in press). A quantitative analysis of the reasons for VCR penetration worldwide. In J. Bryant & J. Salvaggio (Eds.), *Media use in the information age: Emerging patterns of adoption and consumer use*. Hillsdale, NJ: Erlbaum.

Straubhaar, J. D., & Viscasillas, G. M. (1987). *The impact of pirated English-language cable TV in the Dominican Republic*. Unpublished paper, Department of Telecommunication, Michigan State University, East Lansing, MI.

Suraiya, B. (1983, October 6). India's dream merchants face up to a nightmare. *Far Eastern Economic Review*, pp. 80–81.

Suryadinata, L. (1985, April 4). Videos and books battle for the Kung-fu audience. *Far Eastern Economic Review*, pp. 38–39.

Tan Seok Yan. (1987). *Broadcasting system in Malaysia—The influence of videocassette recorders*. East Lansing, MI: Michigan State University, Department of Communication.

Thieves who steal your TV shows. (1980, September 9). *Nottingham Evening Post*, p. 9-C.

Thomas, T. (1983). Jamaica: The Miami connection. *Intermedia, 11*, 57.

Tunstall, J. (1977). *The media are America*. New York: Columbia University Press.

UNESCO. (1949). *Press, film, radio*. Paris: UNESCO.

UNESCO. (1980). *UNESCO statistical yearbook*. Paris: UNESCO.

UNESCO. (1981). *UNESCO statistical yearbook*. Paris: UNESCO.

UNESCO. (1985). *UNESCO statistical yearbook*. Paris: UNESCO.

USAFE television story. (1955, December 13). Washington, D.C.: U.S. Department of Defense.

USIA (United States Information Agency). (1973, August 30). *Media habits of priority groups in Saudi Arabia.* Research report R-20-73A.

USIA (United States Information Agency). (1981). *Media use by the better educated in major Mexican cities.* United States Information Agency, Office of Research, Washington, D.C.

USIA (United States Information Agency). (1982). *Media use by the better educated in four Brazilian cities.* United States Information Agency, Office of Research, Washington, D.C.

Varis, T. (1984). The international flow of television programs. *Journal of Communication, 34*(1), 143–152.

Varis, T. (1974). Global traffic in television. *Journal of Communication, 24*(1), 102–109.

Varis, T. & Nordenstreng, K. (1973). *Global Traffic in Television.* Paris: UNESCO Reports and Papers on Mass Communication.

Veja. (1981, July 22), pp. 41, 64.

Velarde, E. G. (1986, August 10). How video shortchanges film trade. *The Tribune* (Manila), p. 15.

Via, S. C. da (1977). *Televisão e consciência de classe.* Petrópolis, Brazil: Vozes.

Victor Company of Japan. (n.d.). *JVC HR-7600MS.* Tokyo, Japan: JVC.

Video: A media revolution? (1985). *Communication Research Trends, 6*(1).

Video boom. (1983, October 6). *India Today,* pp. 54–60.

Videocassete no Brasil: a maquina do ano. (1986, December 24). *Veja,* pp. 54–61.

Video cassette recorders: National figures. (1983). *Intermedia, 11,* 39.

Video challenge. (1984, May 4). *Asiaweek,* p. 43.

"Video piracy": The other side. (1984, November 27). *Daily News* (Colombo).

Video pirate scourge. (1982, November 12). *Asiaweek,* pp. 46–47.

Wang, G. (1986). Video boom in Taiwan: Blessings or curse? *The Third Channel, 2*(1), 365–377.

Watts, D. (1985, March 2). Two-tape video irks Hollywood. *The Times* (London).

Wells, A. (1972). *Picture tube imperialism?: The impact of U.S. television on Latin America.* Maryknoll, NY: Orbis Books.

Wetzel, H. (1986). *Telecommunications equipment markets in the Caribbean/ Central American Area.* Washington, D.C.: United States Department of Commerce.

White, R. (1976). *An alternative pattern of basic education: Radio Santa Maria* (Experiments and innovations in education 30). Paris: UNESCO, Institute for Education.

Wiarda, H. (1981). *Corporatism and national development in Latin America.* Boulder, CO: Westview Press.

Wikan, U. (1985). Living conditions among Cairo's poor—A view from below. *Middle East Journal, 39*(1), 7–26.

World Bank. (1980). *1980 World Bank atlas*. Washington, D.C.

World Bank. (1985). *The World Bank atlas 1985*. Washington, D.C.

World Bank. (1986). *World development report 1985*. New York: Oxford University Press.

Wyman, T. H. (1983). *Trade barriers to United States motion picture and television, prerecorded entertainment, publishing and advertising industries*. New York: CBS.

Yadava, J. S. (1986). *Communication technologies and developing countries: Effects of television and video in India*. Paper presented at International Association for Mass Communication Research meeting, New Delhi, August 25–29.

Ziff, R. (1979, May). Magnetic tape's impact on broadcasting. *Broadcast Engineering*, p. 82.

PERSONAL COMMUNICATION

In the preparation of this volume, the views of the following individuals were solicited by the authors through interviews or correspondence.

Arab World

Hamarneh, Michael, Undersecretary, Ministry of Information, Amman, Jordan, March 17, 1985.

Isa bin Rashid al-Khalifa, Undersecretary, Ministry of Information, Manama, Bahrain, August 26, 1984.

Jahmani, Issa, Director, Press and Publications Office, Ministry of Information, Amman, Jordan, March 17, 1985.

Kheir, M., Jordanian TV program director, Amman, Jordan, January 14, 1984.

Mytton, Graham, head, BBC External Service, International Broadcasting and Audience Research, London, England, January 23, 1985.

Rachty, Jehan, faculty of mass communication, Cairo University, Cairo, Egypt, January 4, 1987.

Raffoul, Sami, manager, Pan Arab Research Center, Kuwait, February 9, 1985.

Shaheen, Yousef, Cairo, Egypt, January 5, 1987.

Seikaly, Samir, owner, Philadelphia Cinema, Amman, Jordan, March 18, 1985.

Wafai, Mohamed, faculty of Mass Communication, Cairo University, Cairo, Egypt, January 4, 1987.

Asia

Ariyaratne, A. T., President, Lanka Jathika Sarvodaya (Sri Lanka), in Singapore, November 21, 1986.

Avellana, Lamberto, owner and director, Documentary Inc., Quezon City, Philippines, August 16, 1986.

Belbase, Subhadra, program director, Worldview International Foundation Nepal, in Singapore, November 21, 1986.

Bernal, Ishmael, film director, Quezon City, Philippines, August 16, 1986.

Brocka, Lino, film director, Quezon City, Philippines, August 21, 1986.

Chan, David, vice president, production, Golden Communications, Hong Kong, August 13, 1986.

Chang King-Yuh, director-general, Government Information Office, Taipei, Republic of China, August 6, 1986.

Chao, Benny C. P., producer, director of production, Central Motion Picture Company, Taipei, Republic of China, August 5, 1986.

Chiang Tsou-Ming, director, Department of Motion Pictures, Government Information Office, Republic of China, August 5, 1986.

Chou Ling-Kong, director, Fee Tang Motion Picture Company, Taipei, Republic of China, August 6, 1986.

Cinco, Fyke, film director, Quezon City, Philippines, August 16, 1986.

Fan, Frank S. L., general manager, Metro-Goldwyn-Mayer of China, Taipei, Republic of China, August 5, 1986.

Fjortoft, Arne, director, Worldview International Foundation, in Karachi, Pakistan, July 7, 1984.

Fong, Allen, movie director, Hong Kong, August 11, 1986.

Gallaga, Peque, film director, Quezon City, Philippines, August 20, 1986.

Goonsekera, Anura, director-general, Rupahavini (TV) Corporation of Sri Lanka, in Singapore, November 20, 1986.

Hu, Joe-Yang, president, Empire Audio Visual Materials and Hwa Kong Enterprises, Taipei, Republic of China, August 6, 1986.

Jen Wan, film director, Taipei, Republic of China, August 7, 1986.

Krishnan, Dato L., managing director, Gaya Filems, Kuala Lumpur, Malaysia, November 17, 1986.

Kuang, Sunshine, director, Department of Radio and TV Affairs, Government Information Office, Taipei, Republic of China, August 8, 1986.

Lacaba, José F., screenwriter, Quezon City, Philippines, August 22, 1986.

Lam, Peter, distribution manager, Cinema City Co. Ltd., Hong Kong, August 12, 1986.

Lan Tsu-Wei, reporter, *United Daily News*, Taipei, Republic of China, August 7, 1986.

Liu Shou-chi, section chief, Department of Motion Pictures, Government Information Office, Taipei, Republic of China, August 5, 1986.

Abbas Muzaffar, director, Pakistan Information Service Academy, Lahore, Pakistan, July 12, 1984.

Nizami, Arif, executive editor, *Nawa-i-Waqt*, Lahore, Pakistan, July 9, 1984.

Rao, Leela, professor of communications, University of Bangalore (India), in Singapore, November 23, 1986.

Syed Mumtaz Saeed, director, National Management Development Center, Karachi, Pakistan, July 18, 1984.

Tchii, Danny, owner, ESC Inc., Taipei, Republic of China, August 5, 1986.

Tu, Richard C., managing director, China Educational Recreation Ltd., Taipei, Republic of China, August 4, 1986.

Yadava, Y. S., head of research, Indian Institute of Mass Communication (New Delhi), in Prague, Czechoslovakia, August 29, 1984.

Latin America

Aguirre, J. M., professor, Universidad Central de Venezuela, Centro de Comunicación Social, Caracas, Venezuela, June 18, 1985.

Berdegue, Carlos, television producer, personal correspondence from Caracas, Venezuela, December 10, 1984.

Bibliowicz, Azriel, professor, Universidad Javeriana, Bogotá, Colombia, October 1984.

Bisbal, Marcelino, professor, Universidad Central de Venezuela, Caracas, Venezuela, June 18, 1985.

Blanco, Jorge, independent producer, Santo Domingo, Dominican Republic, December 1984.

Bloch, Rosalee, director, Manchete Video, Rio de Janeiro, Brazil, July 16, 1986.

Cohen, Clemente, Vice-ministro de Información y Turismo, Caracas, Venezuela, June 1985.

Cabrera, José, Secretario de Comunicación Social, Santo Domingo, Dominican Republic, December 1984.

Festa, Regina, professor of journalism, Universidade de São Paulo, Brazil, July 1986.

Journalism seminar, Universidad Católica Madre y Maestra, Santo Domingo, Dominican Republic, December 13–15, 1984.

Machado, A., marketing director, Instituto Brasileiro para Estudios de Opinião Publica e Estatistica, Rio de Janeiro, Brazil, November 1982.

Marcelo, Roberto, director, Globovideo, Rio de Janeiro, Brazil, July 16, 1986.

Milanesi, Luis Augusto, video tape project director, Government of the State of São Paulo, São Paulo, Brazil, July 25, 1986.

Ranucci, Karen, Democracy in Communication, Philadelphia, PA., October 1986.

Santoro, Luis Fernando, professor of journalism, University of São Paulo, São Paulo, Brazil, July 1986.

Wallner, Martha, Democracy in Communication, Philadelphia, PA, October 1986.

Video rental shop operators, Santo Domingo, Dominican Republic, December 1984.

Video rental shop operators, Caracas and Barquisimeto, Venezuela, June 1985.

Video rental shop and video club operators, Rio de Janeiro and São Paulo, Brazil, July 1987.

Caribbean

Ewing, Agnes, secretary, Belize Broadcasting Authority, Belize City, Belize, May 29, 1987

King, Emory, former owner, Tropical Vision, Belize City, Belize, May 29, 1987.

Mai, Amalia, editor, *Belize Times,* Belize City, Belize, May 29, 1987.

Vasquez, Nestor, owner, Tropical Vision and Channel 7, chairman of Belize Telecommunication Authority, board member of Belize Broadcasting Authority, Belize City, Belize, May 28, 1987.

York, Ed, program director, Radio One, Belize City, Belize, May 29, 1987.

Index